Synthesis Lectures on Intelligent Technologies

Series Editor

Janusz Kacprzyk, Systems Research Institute, Polish Academy of Science, Warsaw, Poland

Synthesis Lectures on Intelligent Technologies provides highly interdisciplinary research with the potential to change the fundamental principles of our society. It covers applications such as Intelligent Transportation, Humanoids, Self-Driving Cars, IoT, Ambient Intelligence, Smart Cities, Human-computer Interaction, Computational Intelligence, Industry 4.0, Medical Robotics, or Data Science. Synthesis Lectures on Intelligent Technologies brings together up-to-date resources from trusted authors working around the world in all aspects of Intelligent Systems.

Ardashir Mohammadzadeh ·
Mohammad Hosein Sabzalian · Oscar Castillo ·
Rathinasamy Sakthivel · Fayez F. M. El-Sousy ·
Saleh Mobayen

Neural Networks and Learning Algorithms in MATLAB

Springer

Ardashir Mohammadzadeh
Multidisciplinary Center for Infrastructure Engineering
Shenyang University of Technology
Shenyang, China

Oscar Castillo
Division of Graduate Studies and Research
Tijuana Institute of Technology
Tijuana, Mexico

Fayez F. M. El-Sousy
Department of Electrical Engineering, College of Engineering
Prince Sattam Bin Abdulaziz University
Al Kharj, Saudi Arabia

Mohammad Hosein Sabzalian
LabREI - Smart Grid Laboratory, Department of Systems and Energy, FEEC - School of Electrical and Computer Engineering
University of Campinas
Campinas, Brazil

Rathinasamy Sakthivel
Department of Applied Mathematics
Bharathiar University
Coimbatore, Tamil Nadu, India

Saleh Mobayen
Multidisciplinary Center for Infrastructure Engineering
Shenyang University of Technology
Shenyang, China

ISSN 2731-6912 ISSN 2731-6920 (electronic)
Synthesis Lectures on Intelligent Technologies
ISBN 978-3-031-14573-5 ISBN 978-3-031-14571-1 (eBook)
https://doi.org/10.1007/978-3-031-14571-1

This Springer imprint is published by the registered company Springer Nature Switzerland AG
The registered company address is: Gewerbestrasse 11, 6330 Cham, Switzerland

Preface

This book explains the ins and outs of neural networks in a simple approach with clear examples and simulations in MATLAB. The scripts herein are coded for general purposes to be easily extended to a variety of problems. They are vectorized and optimized to run faster and be applicable to high-dimensional engineering problems.

Shenyang, China — Ardashir Mohammadzadeh
Campinas, Brazil — Mohammad Hosein Sabzalian
Tijuana, Mexico — Oscar Castillo
Coimbatore, India — Rathinasamy Sakthivel
Al Kharj, Saudi Arabia — Fayez F. M. El-Sousy
Shenyang, China — Saleh Mobayen

The original version of the book was revised: Author provided belated name corrections have been incorporated. The correction to the book is available at https://doi.org/10.1007/978-3-031-14571-1_9

Contents

Introduction

1

1.1 Overview

Artificial neural networks are now extensively studied in order to achieve human-like efficiency. These networks consist of some linear and nonlinear computational elements that operate in tandem.

Neural networks are cutting-edge computational systems and methods for machine learning, knowledge representation, and, finally, the application of acquired knowledge to predict outputs from complex systems. The main concept behind these networks is (to some extent) inspired by how the biological neural system processes data and information in order to learn and create knowledge. The main component of this concept is the development of novel structures for information processing systems. This system consists of many extremely interconnected processing elements known as the neurons that cooperate to solve problems and transfer information via synapses (electromagnetic communications). If a cell is damaged in these networks, other cells can compensate for its absence and contribute to its reconstruction. These networks are capable of learning. For instance, by applying burn to touch nerve cells, the cells learn not to approach hot objects, and the system learns to correct its mistake thanks to the algorithm. These systems learn comparatively; in other words, a new input is provided, and the weights of synapses change in a way that the system can generate accurate responses.

There is no agreement among researchers on how to define a neural network; however, most agree that it consists of a network of simple processing elements (neurons) capable of displaying an overall complex behavior determined by the relationship between processing elements and element parameters. The main and inspiring source for this technique is to test the central nervous system and neurons (axons, multiple branches of nerve cells, and junctions of two nerves), which are among the most important components of nervous system information processing. Simple nodes (processing elements) or units are interlinked to form a network of nodes in a neural network model. This is why they are referred to as "neural networks." Although a neural network should not be adaptable in

A. Mohammadzadeh et al., *Neural Networks and Learning Algorithms in MATLAB*, Synthesis Lectures on Intelligent Technologies,
https://doi.org/10.1007/978-3-031-14571-1_1

and of itself, it can be used practically thanks to certain algorithms designed to change the communication weight in a network (to create the desired signal).

A data structure that can function as a neuron can be created through computer programming. The network can then be trained by designing a network of these interconnected artificial neurons, developing a training algorithm for the network, and applying this algorithm to its network.

McCulloch and Pitts, two leading figures in neural networks, conducted research on the interconnection ability of a neuron's model by the end of 1940. They proposed a computational model based on a basic neuron-like element. In 1949, Donald O. Hebb proposed a learning rule to adapt connections between artificial neurons. A few years later, in 1958, Rosenblatt proposed the perceptron algorithm and the statistical separability theory accordingly.

1.2 Some Applications of Neural Networks

1. Pattern Recognition: This application includes facial recognition, fingerprint recognition, voice and speech recognition, handwriting recognition, etc.
   This mechanism, for example, is used in banks to compare signatures while withdrawing cash from an account to the signatures recorded in the account records. This is a fundamental application of neural network chips.
2. Medicine: This application includes electrocardiogram signal analysis and recognition as well as a trained network that can diagnose diseases and even prescribe medications.
3. Commercial Applications: They include any types of business decisions that are difficult to make such as decision-making that requires extensive data of a specific objective. For instance, networks are widely used in the stock exchange market to predict stock fluctuations based on previous data.
4. Artificial Intelligence (AI): Many AI experts believe that artificial neural networks are the best and, in many cases, the only hope of designing a smart machine.
5. Visual data compression for data reduction.
6. Noise elimination on telecommunication lines.
7. Military Systems: This application includes submarine mine detection, elimination of abnormal sounds in radar tracking systems, etc.
   Constructing and Operating Building Structures: Since neural networks process and analyze data at high rates, the time spent determining optimal structure is reduced.
8. Marketing: Networks are used in online advertising to improve and optimize sales.
9. Monitoring: Dangers in spacecraft, for example, can be predicted by analyzing audio levels transmitted from the spacecraft. This method has also been tested in rails for the analysis of the sounds produced by diesel engines.

There are some other applications of neural networks: risk analysis systems, pilot-less aircraft control, welding quality analysis, computer quality analysis, emergency room

(ER) testing, oil and gas exploration, truck braking detection systems, loan risk estimation, spectral recognition, medication detection, industrial control processes, error management, sound recognition, hepatitis diagnosis, remote data retrieval, submarine mine detection, 3D object recognition, handwriting and facial recognition, etc. In general, neural network applications can be classified as follows: correspondence (the network recognizes confusing patterns), clustering, categorization, identification, pattern reconstruction, generalization (obtaining a correct response to an input stimulus that has not previously been trained in the network), and optimization. Nowadays, neural networks are used in a variety of tasks such as pattern recognition, which includes handwriting recognition (HWR), speech recognition, visual recognition, and other similar tasks as well as text and/or image classification. Artificial neural networks are increasingly employed to control or model the systems in which internal structures are unknown or extremely complex. For instance, a neural network can be used in engine input control, in which the neural network learns the control function itself [1].

1.3 Different Types of Neural Network Training

Supervised Learning: This task is performed by focusing on a specific subject and presenting a variety of examples. The network analyzes the input data and examples. After a while, it will be able to recognize a new type of example that it has never seen before [2].

- Unsupervised Learning: A higher level of learning that is less commonly used nowadays.
- Reinforcement Learning: Hidden-Mode Markov Decision Processes (HM-MDPs): The main components of a Markov model are a collection of states, a collection of actions, transitions, and the instantaneous added value of each action.

  Suitable problems for learning neural networks [3]:

- There is an error in the training data. For instance, the training data may contain noise caused by sensor data collected from different devices such as cameras and microphones.
- The objective function has continuous values.
- There is sufficient time for learning. In comparison with other techniques such as decision trees, this method requires more learning time.
- The objective function does not need to be changed, for it is difficult to change the weights learned by the network.

1.4 Learning Principles in Neural Networks

Changes in the weights of links between neurons are used to train (learn) neural networks. There are two approaches to neural network training in general: supervised and unsupervised. The values of data (explanatory variables) and outputs (dependent variables) are introduced into the model in supervised training, and the connection weights are regulated to obtain output values as close to the outputs as possible. However, unsupervised training introduces only the values of data to the model, and the learning steps are taken in the absence of the previously introduced output values (dependent variable) [4, 5].

References

1. N. Karayiannis, A.N. Venetsanopoulos, *Artificial Neural Networks: Learning Algorithms, Performance Evaluation, and Applications*, vol 209 (Springer Science & Business Media, Berlin, 2013)
2. S.S. Haykin, S.S. Haykin, S.S. Haykin, S.S. Haykin, *Neural Networks and Learning Machines*, vol. 3 (Pearson, Upper Saddle River, 2009)
3. انجمن هوش مصنوعی. Available: www.artificial.ir
4. I. Sutskever, O. Vinyals, Q.V. Le, Sequence to sequence learning with neural networks, in *Advances in Neural Information Processing Systems* (2014), pp. 3104–3112
5. R.M. Neal, *Bayesian Learning for Neural Networks*, vol. 118 (Springer Science & Business Media, Berlin, 2012)

Multilayer Perceptron (MLP) Neural Networks 2

The simplest type of neuron modeling is the perceptron. Since it is difficult to analyze several perceptron types in different layers, we will begin examining one perceptron.

According to Fig. 2.1, a perceptron consists a series of external inputs, an internal input called a bias, a threshold, and an output.

Perceptron learning is the process of determining the appropriate values of W. As a result, perceptron learning is defined as the sum of all possible real weight vector values. The following equation determines the output of a perceptron:

$$\begin{cases} 1 \; w_0 + x_1 w_1 + \cdots + x_n w_n > 0 \\ 0 \; w_0 + x_1 w_1 + \cdots + x_n w_n \leq 0 \end{cases} \tag{2.1}$$

A perceptron can only learn examples that can be separated linearly. These are instances that can be completely separated by a hyperplane. Consider the following figure:

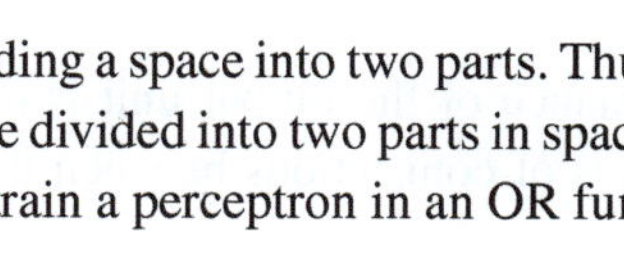

Example: OR Function

A perceptron is capable of dividing a space into two parts. Thus, only functions with positive and negative outputs that can be divided into two parts in space are accurately obtained from a perceptron. We managed to train a perceptron in an OR function as shown in Fig. 2.2.

The layers of a multilayer network are as follows:

A. Mohammadzadeh et al., *Neural Networks and Learning Algorithms in MATLAB*, Synthesis Lectures on Intelligent Technologies,
https://doi.org/10.1007/978-3-031-14571-1_2

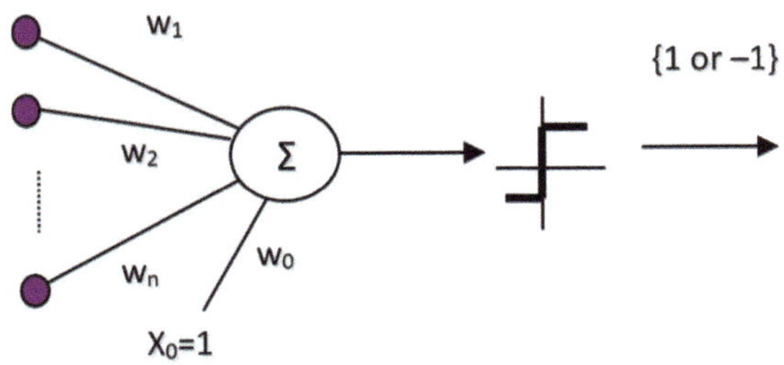

Fig. 2.1 The model of a perceptron

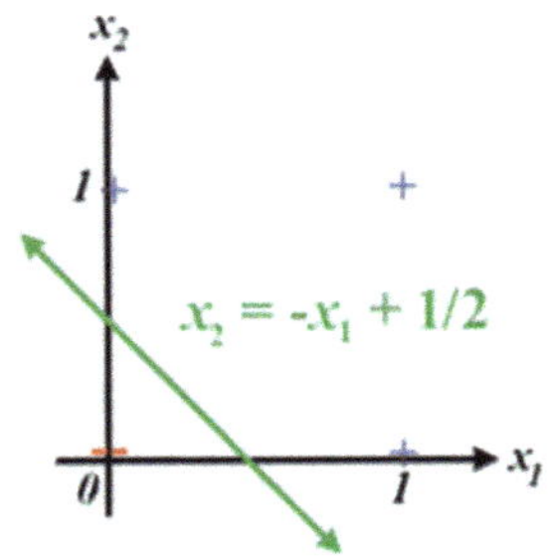

Fig. 2.2 OR function

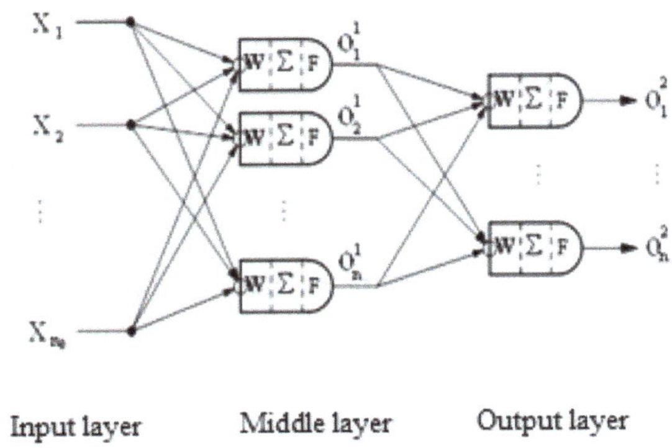

Fig. 2.3 A neural network with three layers

1. Input Layer: It receives data fed into the network in its raw form.
2. Hidden Layers: Inputs and the weight of connections between them and the hidden layers determine the performance of these layers. When a hidden unit should be activated, the weights between input and hidden units are employed to determine when it should be activated.
3. Output Layer: The performance of the output unit is determined by the activity of a hidden layer and the weights of connections between the hidden and output layers.

Figure 2.3 depicts a three-layer neural network.

Where $[x_1, \ldots, x_{n0}]$ denotes the neural network input vectors, and $[o_1^1, \ldots, o_h^1]$ indicates the middle layer outputs, whereas $[o_1^2, \ldots, o_n^2]$ represents the neural network outputs.

If the matrix of weights connected to the middle and input layer neurons is represented by w_1 and the matrix of weights connected to the middle and output layer neurons is represented by w_2, the input vector is represented by x, the activation functions in the middle layer and the output are represented by F_1 and F_2, the output vector of the middle layer is represented by o_1, and the output network vector is represented by o_2, the neural network output is as follows:

$$net_1 = w_1^T x \tag{2.2}$$

$$o_1 = F_1(net_1) \tag{2.3}$$

$$net_2 = w_2^T o_1 \tag{2.4}$$

$$o_2 = F_2(net_2) \tag{2.5}$$

The following MATLAB snippet was written to calculate the neural network output:

Snippet for calculating neural network output

```
function
[o1,o2,net1,net2]=FeedForward_NN(u,b1,b2,w1,w2)
 In this function, b1 and b2 are biases, and w1 and w2 are the middle and
output layer weight matrices.
%% Activation functions
    f1 =@(x)( 8-exp(-x))./(1+exp(-x)); bipolar sigmoid
                                            activation function
%f2 =@(x)( 8-exp(-x))./(1+exp(-x));
    f2 =@(x) 1./(1+exp(-x)); monopolar sigmoid activation
                                                      function
%% Hidden Lay
b1 bias and u neural network input
U1=[b1;u];
net1=w1'*u1;
o1=f1(net1);
%% Out Layer
u2=[b2;o1];
net2=w2'*u2;
o2=f2(net2);
```

2.1 Training Based on Error Backpropagation

The feedforward neural networks use this algorithm proposed by Rumelhart and McClelland in 1986. The term "feedforward" refers to the placement of artificial neurons in successive layers that transmit their outputs (signals) forward. Backpropagation also refers to the process of feeding errors backward in the network to correct the weights and then repeating the input from the forward path to the output [1]. Error backpropagation is a supervised method which means that the input samples are labeled and the expected output of each is known in advance. Thus, the network output is compared with these ideal outputs, and the network error is computed. This algorithm is based on the assumption that the network weights are selected at random first. At each step, the network output is calculated, and the weights are corrected based on the difference between it and the desired output so that the error will be minimized in the end [2].

Figure 2.4 depicts the learning process of a simple neural network. An adaptive formula for synaptic weights in the network is obtained by providing the optimal solution for d so that the network output will be close enough to d or the neural network acquires the necessary knowledge of the optimal response d. A backpropagation solution derived from descenting gradient based on the chain rule can be employed to solve this problem. The details of this procedure are as follows. The square of the instantaneous error between the optimal solution d and the network output in step k is regarded as a standard function of network performance:

$$E(k) = \frac{1}{2}e^2(k) = \frac{1}{2}(d(k) - o_2(k))^2 \tag{2.6}$$

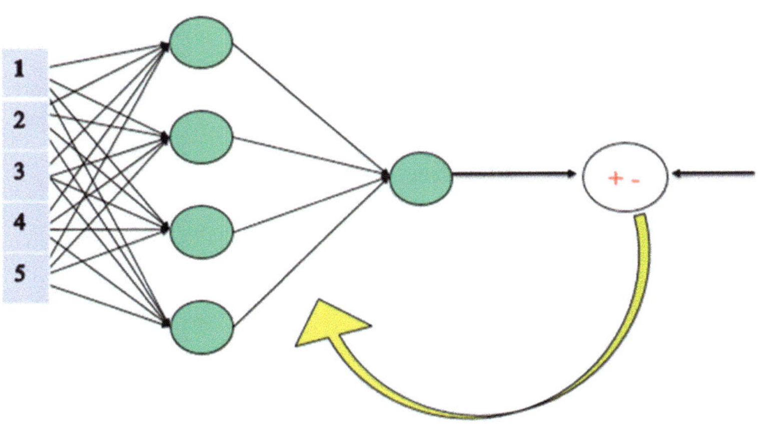

Fig. 2.4 A simple neural network with one output

$$w_2(k+1) = w_2(k) + \left(-\eta_w \cdot \frac{\partial E(k)}{\partial w_2(k)}\right) \tag{2.7}$$

$$\frac{\partial E(k)}{\partial w_2(k)} = \frac{\partial E(k)}{\partial o_2(k)} \frac{\partial o_2(k)}{\partial net_2(k)} \frac{\partial net_2(k)}{\partial w_2(k)} = -eF_2' o_1 \tag{2.8}$$

$$\frac{\partial E(k)}{\partial w_1^j(k)} = \frac{\partial E(k)}{\partial e(k)} \frac{\partial e(k)}{\partial o_2(k)} \frac{\partial o_2(k)}{\partial net_2(k)} \frac{\partial net_2(k)}{\partial o_1^j(k)} \frac{\partial o_1^j(k)}{\partial net_1^j(k)} \frac{\partial net_1^j(k)}{\partial w_1^j(k)} = -eF_2' w_2^j F_1' x \tag{2.9}$$

where η_w refers to the gradient descent training rate, and w_2^j denotes the weight attached to the jth neuron and the output, whereas w_1^j represents the vector of weights connected to the jth neuron in the first layer. F_2' and F_1' refer to the activation function derivatives in the output and first layers, respectively. This training procedure is repeated until the cost function is sufficiently reduced in the error backpropagation algorithm.

2.2 Implementation in MATLAB

Example Use a two-layer MLP neural network employed to estimate the following function.

$$\mathrm{y_p}(\mathrm{k}+1) = \mathrm{f}\left(\mathrm{y_p}(k), y_p(k-1), y_p(k-2), u(k), u(k-1)\right)$$
$$\text{where } f(x_1, x_2, x_3, x_4, x_5) = \frac{x_1 x_2 x_3 x_5 (x_3 - 1) + x_4}{1 + x_2^2 + x_3^3} \tag{2.10}$$

To train a neural network with an arbitrary number of inputs and outputs, the following function was coded in MATLAB:

MLP neural network training function with gradient descent

```
function [w1,w2]=GD_MLP(e,eta,u,o1,net1,net2,w1,w2)
%% Derivative of Activation functions
df1 =@(x)2*exp(-x)./(1+exp(-x)).^2;
                                bipolar sigmoid activation function
df2 =@(x)2*exp(-x)./(1+exp(-x)).^2;

delta2=e.*df2(net2);
w2=w2+eta*o1*delta2';

delta1=(w2*delta2).*df1(net1);
w1=w1+eta*u*(delta1)
```

Nonlinear system estimation snippet (2-10) using neural networks

```
clear all
clc

%%  Generate Training and Test data
u(:,1)= 8-2*rand(1000,1);
y(3,1)=0;
% f(x1,x2,x3,x4,x5)=( x1*x2*x3*x5(x3-1)+x4 ) / (1+x2^2+x3^2)
for t=4:length(u);

  y(t) =(y(t-1)*y(t-2)*y(t-3)*u(t-2)*(y(t-3)-1)+u(t-1))/(1 +
y(t-2)^2+y(t-3)^2);
end

%% Input data
data=[[0;y(1:end-1)] [0;0;y(1:end-2)] [0;0;0;y(1:end-3)]
[0;u(1:end-1)] [0;0;u(1:end-2)] ];

nn=randperm(1000);
Train_data=data(nn(1:700),:);
Target_train=y(nn(1:700));

Test_data=data(nn(701:end),:);
Target_test=y(nn(701:end));

%% NN Initial parameters
num_neuron=4;
w1=rand(6,num_neuron);
w2=rand(num_neuron+1,1);
eta=.5;
b1=1;
b2=1;
%% Train ==================================================
for epoch=1:100
    epoch
    %% Train
    E=0;
    nn=randperm(700);
    kk=1;
    for ii=nn
        x=Train_data(ii,:)';
```

```
        [o1,o2,net1,net2]=FeedForward_NN(x,b1,b2,w1,w2);
        e=Target_train(ii)-o2;

[w1,w2]=GD_MLP(e,eta,[b1;x],[b2;o1],net1,net2,w1,w2);
        E=E+e^2;
    end
   E/700
   MSE_Train(epoch)= E/700;

   %% Test
     E=0;
    for ii=1:300
      x=Test_data(ii,:)';
      [o1,o2,net1,net2]=FeedForward_NN(x,1,1,w1,w2);
      e=Target_test(ii)-o2;
      E=E+e^2;
      yest(ii)=o2;
    end
    E/300
   MSE_Test(epoch)=E/300;
end
%% plot MSE
%%
  plot(MSE_Train(1:end),'--b','linewidth',2)
  hold on
  plot( MSE_Test,'--r','linewidth',2)
%% Test
============================================================
=====
    E=0;
for ii=1:300
   x=Test_data(ii,:)';
   [o1,o2,net1,net2]=FeedForward_NN(x,1,1,w1,w2);
   e=Target_test(ii)-o2;
   E=E+e^2;
   yest(ii)=o2;
end
  E/300

 %% -------------- plot out -----------------------------
  plot(yest,'--b','linewidth',2)
  hold on
  plot(Target_test,'--r','linewidth',2)
```

Figure 2.5 demonstrates the output and the estimated output diagrams through the neural network. Accordingly, the neural network performed properly in estimating the nonlinear system depicted in Fig. 2.5

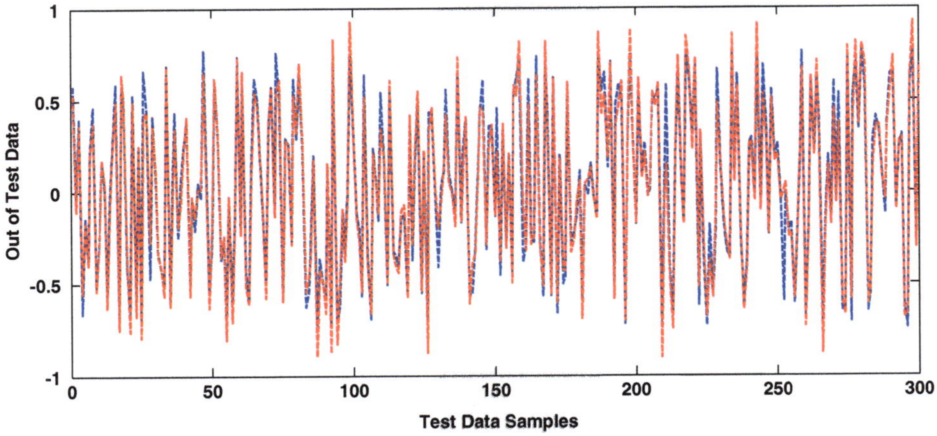

Fig. 2.5 The system output and the diagram estimated through a neural network

2.3 Application of Neural Network in Classification

Classification is among the most important applications of neural networks. A simulation example is provided to help better understand the capabilities of learning algorithms and neural networks. Data are collected from ten different signals, and the neural network is trained to separate these signals. Each signal has 23 features, and there are 1024 feature vectors for training and 400 attribute vectors for testing and evaluating each signal. You can either download the data of the third tutorial session at the website[1] or use any types of data in this script for training and testing. All you have to do is enter training and testing data according to the script and the problem dimensions you want.

[1] https://simref.org/nn_matlab/.

MATLAB classification script with the desired number of inputs and outputs

```
clear all
clc



%%  Load data

load data_classification

%% Input data

Train_data=reshape(Xtrain,size(Xtrain,1),size(Xtrain,2)*size
(Xtrain,3));
Target_train=repmat([1:10],size(Xtrain,2),1);
Target_train=reshape(Target_train,numel(Target_train),1);

%----------------------------
Test_data=reshape(Xtest,size(Xtest,1),size(Xtest,2)*size(Xte
st,3));
Target_test=repmat([1:10],size(Xtest,2),1);
Target_test=reshape(Target_test,numel(Target_test),1);


%% NN Initial parameters
num_neuron=4;
w1=rand(24,num_neuron);
w2=rand(num_neuron+1,10);
eta=.09;
b1=1;
b2=1;
O_target=zeros(10,1);
%% Train ====================================================
for epoch=1:400
    epoch
    %% Train
    E=0;
    nn=randperm(size(Train_data,2));

    for ii=nn
        x=Train_data(:,ii);
        [o1,o2,net1,net2]=FeedForward_NN(x,b1,b2,w1,w2);
        O_target=zeros(10,1);
        O_target(Target_train(ii))=1;
        e=O_target-o2;
```

```
[w1,w2]=GD_MLP(e,eta,[b1;x],[b2;o1],net1,net2,w1,w2);
        E=E+norm(e)^2;
    end
   E/size(Train_data,2)
   MSE_Train(epoch)= E/size(Train_data,2);

   %% Test
     E=0;
    for ii=1:size(Test_data,2)
      x=Test_data(:,ii);
      [o1,o2,net1,net2]=FeedForward_NN(x,b1,b2,w1,w2);
      O_target=zeros(10,1);
      O_target(Target_test(ii))=1;
      e=O_target-o2;
      E=E+norm(e)^2;
    end
    E/size(Test_data,2)
   MSE_Test(epoch)=E/size(Test_data,2);
end
%% plot MSE
%%
  plot(MSE_Train(1:end),'--b','linewidth',2)
  hold on
  plot( MSE_Test,'--r','linewidth',2)
%% Test
==============================================================
=====
          Assessing the accuracy of neural network detection
The test data is fed into the neural network one by one, and
the neural network output is checked in a 10 × 10 matrix
whose dimensions correspond to the number of outputs. Assume
the first data is for the fifth signal; neural network
outputs are obtained for this data, and it is then examined
to determine which neural network output is closest to 1 (or
is greater than the others). Assume the eighth output is
greater than the others. In this case, it is assumed that
the neural network recognizes this data as corresponding to
the eighth signal, and one unit is added to the eighth line
in matrix C's fifth column. If the neural network detected
this data as corresponding to, for example, the tenth
signal, a unit would be added to the tenth line in the fifth
columns.

C=zeros(10,10);
for ii=1:size(Test_data,2)
      x=Test_data(:,ii);
      [o1,o2,net1,net2]=FeedForward_NN(x,b1,b2,w1,w2);
      O_target=zeros(10,1);
      O_target(Target_test(ii))=1;
      [a,b]=max(abs(o2));
      C(b,Target_test(ii))=C(b,Target_test(ii))+1;
end
C
sum(sum(C.*eye(10)))*100/size(Test_data,2)
```

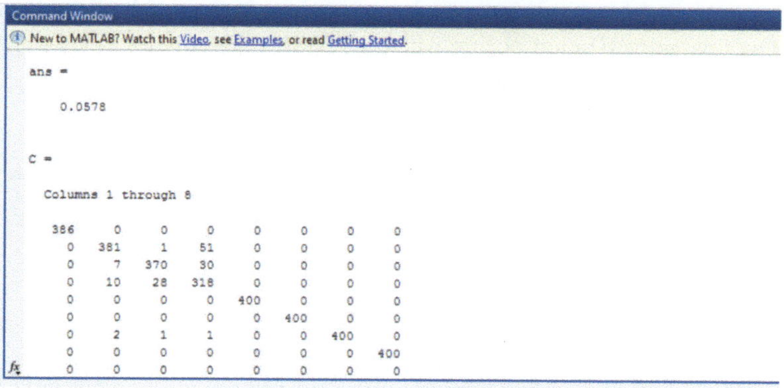

Fig. 2.6 The output diagram in the classification problem

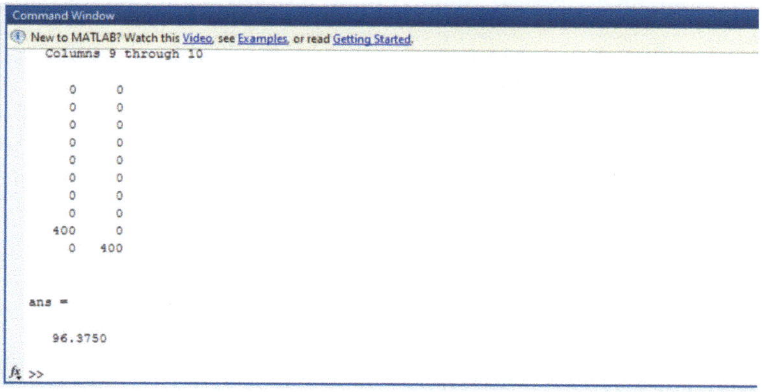

Fig. 2.7 The diagram of detection accuracy in the classification problem

Figure 2.6 depicts the output diagram. After 400 iterations, the MSE value reached 0.0578. Figure 2.7 depicts the detection accuracy diagram. Accordingly, the evaluation accuracy reached 96%.

2.4 Over-Parameterization

Over-parameterization, or the search for optimal neurons and structures, is an important consideration while training neural networks. In other words, the number of adjustable parameters in a neural network must be proportional to the amount of training data in order for the network to be properly trained. Over-parameterization occurs when there are too many neural network parameters and the network is not well trained. The previous

MATLAB script is run for the various number of neurons in the middle layer to clarify this problem. Over-parameterization occurs when the MSE diagrams for testing and training data are separated and the MSE test starts to increase.

MATLAB script for detecting the optimal number of middle layer neurons to avoid over-parameterization

```
clear all
clc
%%  Load data
load data_classification
%% Input data
Train_data=reshape(Xtrain,size(Xtrain,1),size(Xtrain,2)*size
(Xtrain,3));
Target_train=repmat([1:10],size(Xtrain,2),1);
Target_train=reshape(Target_train,numel(Target_train),1);
%----------------------------
Test_data=reshape(Xtest,size(Xtest,1),size(Xtest,2)*size(Xte
st,3));
Target_test=repmat([1:10],size(Xtest,2),1);
Target_test=reshape(Target_test,numel(Target_test),1);

%% NN Initial parameters
kk=1;
for k=[4 8 12 16 20 24 28 32]
num_neuron=k;
w1=rand(24,num_neuron);
w2=rand(num_neuron+1,10);
eta=.09;
b1=1;
b2=1;
O_target=zeros(10,1);
%% Train ===================================================

for epoch=1:20
    epoch
    %% Train
    E=0;
    nn=randperm(size(Train_data,2));

    for ii=nn
        x=Train_data(:,ii);
        [o1,o2,net1,net2]=FeedForward_NN(x,b1,b2,w1,w2);
        O_target=zeros(10,1);
        O_target(Target_train(ii))=1;
        e=O_target-o2;

[w1,w2]=GD_MLP(e,eta,[b1;x],[b2;o1],net1,net2,w1,w2);
        E=E+norm(e)^2;
    end
   E/size(Train_data,2)
```

```
    %% Test
%       E=0;
%      for ii=1:size(Test_data,2)
%        x=Test_data(:,ii);
%        [o1,o2,net1,net2]=FeedForward_NN(x,b1,b2,w1,w2);
%        O_target=zeros(10,1);
%        O_target(Target_test(ii))=1;
%        e=O_target-o2;
%        E=E+norm(e)^2;
%      end
%      E/size(Test_data,2)
%     MSE_Test(epoch)=E/size(Test_data,2);
end%epoch

MSE_Train(kk)= E/size(Train_data,2);
kk=kk+1;
end%k
%% plot MSE
%%
  plot(MSE_Train(1:end),'--b','linewidth',2)
  hold on
  plot( MSE_Test,'--r','linewidth',2)
%% Test
============================================================
=====

C=zeros(10,10);
for ii=1:size(Test_data,2)
      x=Test_data(:,ii);
      [o1,o2,net1,net2]=FeedForward_NN(x,b1,b2,w1,w2);
      O_target=zeros(10,1);
      O_target(Target_test(ii))=1;
      [a,b]=max(abs(o2));
      C(b,Target_test(ii))=C(b,Target_test(ii))+1;
end
C
sum(sum(C.*eye(10)))*100/size(Test_data,2)
```

The number of iterations in this snippet was set to 20 in order to increase the speed with which the snippet ran. You can obtain an accurate solution by adding up the number of iterations. Figure 2.8 depicts the output script. Accordingly, when the number of neurons reaches 20, over-parameterization occurs. If the number of middle layer neurons is set to exceed 20, you will notice that no training occurs.

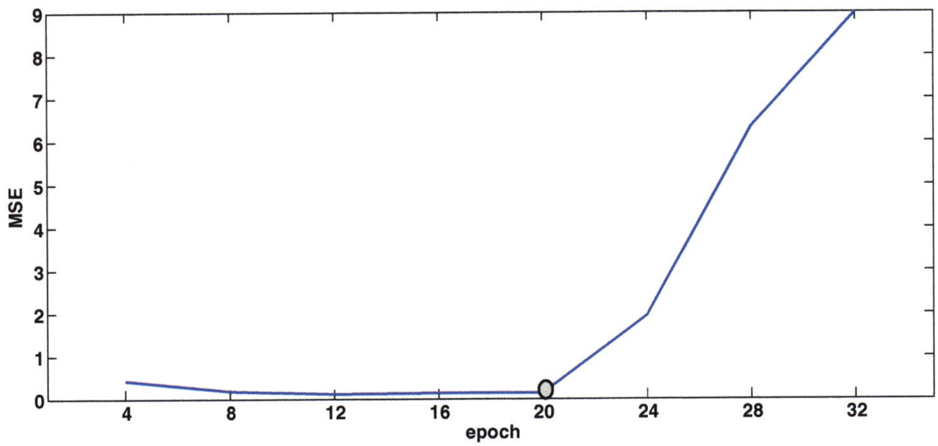

Fig. 2.8 The optimal number of middle-layer neurons and analysis of over-parameterization

2.5 Over-Training

Another important point in neural network training is over-training. If there are too many training iterations, the neural network depends on training data and cannot detect test data accurately. In other words, the number of optimal iterations should be determined, and testing should be terminated once the optimal iterations are found. For this purpose, the training process is repeated in the number of iterations specified, and MSE diagrams of testing and training data are drawn. The optimal number of iterations is the point at which the MSE diagrams of testing and training are separated from each other and the MSE of test data begins to increase (over-training occurs at this point).

2.6 Training Based on Full Propagation

At this point, unlike the previous technique, the weights of the neural network are not updated; however, the weights are updated only once per iteration. In other words, weight changes are saved for each dataset before being applied all at once. For this purpose, the following MATLAB script is presented as below:

MATLAB classification script through full propagation

```
clear all
clc
 %%  Load data
load data_classification
 %% Input data

Train_data=reshape(Xtrain,size(Xtrain,1),size(Xtrain,2)
*size(Xtrain,3));
Target_train=repmat([1:10],size(Xtrain,2),1);
Target_train=reshape(Target_train,numel(Target_train),1
);

%----------------------------
Test_data=reshape(Xtest,size(Xtest,1),size(Xtest,2)*siz
e(Xtest,3));
Target_test=repmat([1:10],size(Xtest,2),1);
Target_test=reshape(Target_test,numel(Target_test),1);


%% NN Initial parameters
num_neuron=20;
w1=rand(24,num_neuron);
w2=rand(num_neuron+1,10);
eta=.09;
b1=1;
b2=1;
O_target=zeros(10,1);
%% Train
=====================================================
```

```
delta_w1=zeros(size(w1));
delta_w2=zeros(size(w2));
max_deta2=ones(size(delta_w2));
max_deta1=ones(size(delta_w1));

for epoch=1:400
    epoch
    %% Train
    E=0;
    nn=randperm(size(Train_data,2));

    delta_w1=zeros(size(w1));
    delta_w2=zeros(size(w2));
    for ii=nn
        x=Train_data(:,ii);
[o1,o2,net1,net2]=FeedForward_NN(x,b1,b2,w1,w2);
        O_target=zeros(10,1);
        O_target(Target_train(ii))=1;
        e=O_target-o2;
[dw1,dw2]=GD_MLP_2(e,1,[b1;x],[b2;o1],net1,net2,w1,w2);

        delta_w2=delta_w2+dw2;
        delta_w1=delta_w1+dw1;
        E=E+norm(e)^2;
    end

    delta_w1(delta_w1>.10)=.10;
    delta_w2(delta_w2>.10)=.10;
    delta_w1(delta_w1<-.10)=-.10;
    delta_w2(delta_w2<-.10)=-.10;

    w1=w1+eta*delta_w1;
    w2=w2+eta*delta_w2;
 E/size(Train_data,2)
MSE_Train(epoch)= E/size(Train_data,2);

   %% Test
%      E=0;
%     for ii=1:size(Test_data,2)
%       x=Test_data(:,ii);
%
[o1,o2,net1,net2]=FeedForward_NN(x,b1,b2,w1,w2);
%       O_target=zeros(10,1);
%       O_target(Target_test(ii))=1;
%       e=O_target-o2;
%       E=E+norm(e)^2;
%     end
%     E/size(Test_data,2)
%    MSE_Test(epoch)=E/size(Test_data,2);
end
```

```
%% plot MSE
%%
%   plot(MSE_Train(1:end),'--b','linewidth',2)
%   hold on
%   plot( MSE_Test,'--r','linewidth',2)
%% Test
=======================================================
==========

C=zeros(10,10);
for ii=1:size(Test_data,2)
      x=Test_data(:,ii);
      [o1,o2,net1,net2]=FeedForward_NN(x,b1,b2,w1,w2);
      O_target=zeros(10,1);
      O_target(Target_test(ii))=1;
      [a,b]=max(abs(o2));
      C(b,Target_test(ii))=C(b,Target_test(ii))+1;
end
C
sum(sum(C.*eye(10)))*100/size(Test_data,2)
```

References

1. X. Yu, M.O. Efe, O. Kaynak, A general backpropagation algorithm for feedforward neural networks learning. IEEE Trans. Neural Networks **13**, 258–254 (2002)
2. R.M. Neal, *Bayesian Learning for Neural Networks*, vol. 118 (Springer Science & Business Media, Berlin, 2012)

3 Neural Networks Training Based on Recursive Least Squares (RLS)

3.1 RLS Training Technique

To train neural networks through the RLS method, the derivatives of the neural network outputs compared to the adjustable parameters should be obtained in a manner similar to the gradient descent. The following cost function should be minimized in order to regulate the output weights and the middle layer of an MLP neural network based on RLS:

$$E(k) = \frac{1}{2}\left[d(k) - o_o^2(k)\right]^2 \tag{3.1}$$

where $d(k)$ denotes an optimal output, and $o^2(k)$ represents the real output. The network errors can be easily derived from error functions compared with the total weighted input, and the weights for the output layer are trained through the following algorithm:

$$w_{hj}(k) = w_{hj}(k-1) + p_h(k)\phi_h(k)\left(e_{hj}(k)\right) \tag{3.2}$$

$$w_{o_1}(k) = w_{o_1}(k-1) + p_o(k)\phi_o(k)(e_{o_1}(k)) \tag{3.3}$$

$$p_h(k) = p_h(k-1)[I - K_h(k)\phi_h(k)] \tag{3.4}$$

$$K_h(k) = \frac{p_h(k-1)\phi_h^T(k)}{1 + \phi_h(k)p_h(k-1)\phi_h^T(k)} \tag{3.5}$$

$$p_o(k) = p_o(k-1)[I - K_o(k)\phi_o(k)] \tag{3.6}$$

A. Mohammadzadeh et al., *Neural Networks and Learning Algorithms in MATLAB*,
Synthesis Lectures on Intelligent Technologies,
https://doi.org/10.1007/978-3-031-14571-1_3

$$K_o(k) = \frac{p_o(k-1)\phi_o^T(k)}{1 + \phi_o(k)p_o(k-1)\phi_o^T(k)} \tag{3.7}$$

where w_{hj} represents the vector of the middle layer weights connected to the jth neuron, and w_{o1} denotes the vector of the output layers. Moreover, p_h represents the middle layer covariance matrix, and p_o refers to the output layer covariance matrix. Finally, K_h and K_o indicate the regulate interest.

3.2 Implementation in MATLAB

Consider the classification example from the previous chapter. The neural network parameters are regulated through RLS in this case.

Classification example using RLS

```
clear all
clc
%%  Load data
load data_classification
%% Input data
Train_data=reshape(Xtrain,size(Xtrain,1),size(Xtrain,2)*size
(Xtrain,3));
Target_train=repmat([1:10],size(Xtrain,2),1);
Target_train=reshape(Target_train,numel(Target_train),1);
%----------------------------
Test_data=reshape(Xtest,size(Xtest,1),size(Xtest,2)*size(Xte
st,3));
Target_test=repmat([1:10],size(Xtest,2),1);
Target_test=reshape(Target_test,numel(Target_test),1);
%% NN Initial parameters
num_neuron=4;
w1=rand(24,num_neuron);
w2=rand(num_neuron+1,10);
P1=2e3*eye(numel(w1));
P2=2e1*repmat(eye(num_neuron+1),1,1,10);
b1=1;
b2=1;
O_target=zeros(10,1);
%% Train =====================================================
for epoch=1:40
    epoch
    %% Train
    E=0;
    nn=randperm(size(Train_data,2));

    for ii=nn
```

```
        x=Train_data(:,ii);
        [o1,o2,net1,net2]=FeedForward_NN(x,b1,b2,w1,w2);
        O_target=zeros(10,1);
        O_target(Target_train(ii))=1;
        e=O_target-o2;

[dO_dw1,dO_dw2,dO_dw12,dO_dw22]=dO_dw(e,[b1;x],[b2;o1],net1,
net2,w1,w2);

        for k=1:10

[w2(:,k),P2(:,:,k)]=RLS2(e(k),w2(:,k),P2(:,:,k),dO_dw2(:,k))
;
        end

        E=E+norm(e)^2;
    end
   E/size(Train_data,2)
   MSE_Train(epoch)= E/size(Train_data,2);

   %% Test
     E=0;
    for ii=1:size(Test_data,2)
      x=Test_data(:,ii);
      [o1,o2,net1,net2]=FeedForward_NN(x,b1,b2,w1,w2);
      O_target=zeros(10,1);
      O_target(Target_test(ii))=1;
      e=O_target-o2;
      E=E+norm(e)^2;
    end
    E/size(Test_data,2)
   MSE_Test(epoch)=E/size(Test_data,2);
end
%% plot MSE
  plot(MSE_Train(1:end),'--b','linewidth',2)
  hold on
  plot( MSE_Test,'--r','linewidth',2)
%% Test
============================================================
=====
C=zeros(10,10);
for ii=1:size(Test_data,2)
      x=Test_data(:,ii);
      [o1,o2,net1,net2]=FeedForward_NN(x,b1,b2,w1,w2);
      O_target=zeros(10,1);
      O_target(Target_test(ii))=1;
      [a,b]=max(abs(o2));
      C(b,Target_test(ii))=C(b,Target_test(ii))+1;
end
C
sum(sum(C.*eye(10)))*100/size(Test_data,2)
```

RLS snippet

```
function [w,P]=RLS2(e,w,P,phi2)
    landa=1;
    k=P*phi2*((landa+phi2'*P*phi2)\1);
    P=(eye(length(w))-k*phi2')/landa*P;
    w=w+k*e;
```

Calculating the derivative of the MLP neural network output in relation to the middle and output layer weights

```
function
[dO_dw1,dO_dw2,dO_dw12,dO_dw22]=dO_dw(e,u,o1,net1,net2,w1,w2
)

%% Activation functions
%f1 =@(x)( 8-exp(-x))./(1+exp(-x));
%f2 =@(x)( 8-exp(-x))./(1+exp(-x));
%% Derivative of Activation functions
df1 =@(x)2*exp(-x)./(1+exp(-x)).^2;
%df2 =@(x)2*exp(-x)./(1+exp(-x)).^2;
df2 =@(x)1*exp(-x)./(1+exp(-x)).^2;
delta2=df2(net2);
dO_dw2=(o1*delta2');

delta22=e.*delta2;
dO_dw22=(o1*delta22');

delta1=(w2(2:end,:)*delta2).*df1(net1);
dO_dw1=(u*(delta1)');

delta12=(w2(2:end,:)*delta22).*df1(net1);
dO_dw12=(u*(delta12)');
```

It is preferable to use the following script if all weights are to be trained at the same time:

```
clear all
clc
%%  Load data
load data_classification
%% Input data
Train_data=reshape(Xtrain,size(Xtrain,1),size(Xtrain,2)*size
(Xtrain,3));
Target_train=repmat([1:10],size(Xtrain,2),1);
Target_train=reshape(Target_train,numel(Target_train),1);
%----------------------------
Test_data=reshape(Xtest,size(Xtest,1),size(Xtest,2)*size(Xte
st,3));
Target_test=repmat([1:10],size(Xtest,2),1);
Target_test=reshape(Target_test,numel(Target_test),1);
%% NN Initial parameters
num_neuron=20;
w1=rand(24,num_neuron);
w2=rand(num_neuron+1,10);
P1=1e-2*eye(numel(w1));
P2=1e1*eye(numel(w2));
b1=1;
b2=1;
O_target=zeros(10,1);
rmse_min=100;
%% Train ====================================================
for epoch=1:100

    %% Train
    E=0;
    nn=randperm(size(Train_data,2));

    for ii=nn
        x=Train_data(:,ii);
        [o1,o2,net1,net2]=FeedForward_NN(x,b1,b2,w1,w2);
        O_target=zeros(10,1);
        O_target(Target_train(ii))=1;
        e=O_target-o2;

[dO_dw1,dO_dw2,dO_dw12,dO_dw22]=dO_dw(e,[b1;x],[b2;o1],net1,
net2,w1,w2);

        [w22,P2]=RLS(reshape(w2,numel(w2),1),P2,...

reshape(dO_dw2,numel(dO_dw2),1),reshape(dO_dw22,numel(dO_dw2
2),1));
        [w11,P1]=RLS(reshape(w1,numel(w1),1),P1,...

reshape(dO_dw1,numel(dO_dw1),1),reshape(dO_dw12,numel(dO_dw1
2),1));
        w1=reshape(w11,24,num_neuron);
        w2=reshape(w22,num_neuron+1,10);
```

```
%[w1,foo]=GD_MLP(e,0.09,[b1;x],[b2;o1],net1,net2,w1,w2);
        E=E+norm(e)^2;
    end
    epoch
  % E/size(Train_data,2)
  % MSE_Train(epoch)= E/size(Train_data,2);

   %% Test
     E=0;
    for ii=1:size(Test_data,2)
      x=Test_data(:,ii);
      [o1,o2,net1,net2]=FeedForward_NN(x,b1,b2,w1,w2);
      O_target=zeros(10,1);
      O_target(Target_test(ii))=1;
      e=O_target-o2;
      E=E+norm(e)^2;
    end
   rmse=E/size(Test_data,2)
   %MSE_Test(epoch)=E/size(Test_data,2);

   if rmse_min>rmse
     rmse_min=rmse;
     w1_best=w1;
     w2_best=w2;
   end
end
%% Test
============================================================
=====

C=zeros(10,10);
for ii=1:size(Test_data,2)
      x=Test_data(:,ii);
      [o1,o2,net1,net2]=FeedForward_NN(x,b1,b2,w1,w2);
      O_target=zeros(10,1);
      O_target(Target_test(ii))=1;
      [a,b]=max(abs(o2));
      C(b,Target_test(ii))=C(b,Target_test(ii))+1;
end
C
sum(sum(C.*eye(10)))*100/size(Test_data,2)
```

RLS snippet to train MLP neural network weights together

```
function [w,P]=RLS(w,P,phi,phi2)
    landa=1;
    k=P*phi*((landa+phi'*P*phi)\1);
    P=(eye(length(w))-k*phi')/landa*P;
    w=w+P*phi2;
```

3.3 Comparison with Gradient Descent

If the parameter regulation rule is used in the gradient descent algorithm, then:

$$\begin{aligned} w_2(k+1) &= w_2(k) + \eta\left(F_2' o_1\right)e \\ w_1^j(k+1) &= w_1^j(k+1) + \eta\left(F_2' w_2^j F_1' x\right)e \end{aligned} \tag{3.8}$$

where w_2 is the output layer's weight vector and w_1^j is the weight vector of the jth neuron connected between the input and middle layers.

A comparison of relation 3.8 with relations 3.2 and 3.3 demonstrates that the RLS training rule is the same gradient as the adaptive training rate.

4 Neural Networks Training Based on Second-Order Optimization Technique

4.1 Introduction

In the previous chapters, only the first-order derivative was employed to obtain the rule for regulating neural network parameters. In other words, only the first-order derivative was used to approximate the cost function through Taylor series, whereas higher-order derivatives were overlooked.

Consider the following cost function $E(w)$ based on a weight vector:

$$E(w) = E(w_0) + g^T \Delta w + \frac{1}{2}(\Delta w)^T H \Delta w \tag{4.1}$$

where g denotes the Jacobian matrix, and H refers to the Hessian matrix. To obtain the smallest $E(w)$, it should be derived and set it to zero as follows:

$$\begin{aligned} &\frac{\partial E(w)}{\partial w} = g + H\Delta w = 0 \\ &\Rightarrow \Delta w = H^{-1} g \end{aligned} \tag{4.2}$$

Thus, the inverse of the Hessian matrix must be obtained in the second-order methods. There are several approaches to this end, which will be discussed further below.

A. Mohammadzadeh et al., *Neural Networks and Learning Algorithms in MATLAB*, Synthesis Lectures on Intelligent Technologies,
https://doi.org/10.1007/978-3-031-14571-1_4

4.2 Newton's Method

In Newton's method, H^{-1} is estimated as follows:

$$\lim_{k\to\infty} Q(k) = H^{-1} \tag{4.3}$$

where

$$\begin{aligned} H &= \frac{\partial g}{\partial w} \cong \frac{\Delta g}{\Delta w} \\ \Rightarrow H^{-1} &= \Delta w.\Delta g^{-1} \end{aligned} \tag{4.4}$$

where $\Delta g(k) = g(k) - g(k-1)$

4.3 Levenberg–Marquardt Algorithm

The Hessian matrix is estimated in this algorithm, which is similar to Newton's method. The benefits of this method include its high convergence speed and flexibility. This method falls somewhere between the gradient descent and Newton's method. The Hessian matrix can be written as follows [1]:

$$H = J^T.J \tag{4.5}$$

where

$$g = J^T.e \tag{4.6}$$

And J represents the Jacobian matrix.

In the Levenberg–Marquardt algorithm, the Hessian matrix is estimated as follows:

$$Q(k) = \left[J^T.J + \mu I\right]^{-1} \tag{4.7}$$

As a result of Relation 4.2, the weight regulation rule is as follows:

$$W(k+1) = W(k) - \left(\left[J^T.J + \mu I\right]^{-1}\right).\left(J^T.e\right) \tag{4.8}$$

where μ is a constant number. The Levenberg–Marquardt algorithm approaches Newton's method when the value of μ approaches zero. If a large number is selected for μ, this method approaches gradient descent while maintaining a low training rate.

4.4 Conjugate Gradient (CG) Method

This method is similar to the gradient method, which uses the following parameter regulation rule [2]:

$$\mathrm{w}(\mathrm{k}+1) = \mathrm{w}(\mathrm{k}) + \eta(\mathrm{k})\Delta\mathrm{w}(\mathrm{k}) \tag{4.9}$$

where

$$\Delta\mathrm{w}(\mathrm{k}) = -\mathrm{g}(\mathrm{k}) + \alpha(\mathrm{k}-1)\Delta\mathrm{w}(\mathrm{k}-1) \tag{4.10}$$

One of the methods listed below is used to calculate α:

$$\begin{aligned}
\alpha(\mathrm{k}) &= \frac{(\mathrm{g}(\mathrm{k}+1))^{\mathrm{T}}\mathrm{g}(\mathrm{k}+1)}{(\mathrm{g}(\mathrm{k}))^{\mathrm{T}}\mathrm{g}(\mathrm{k})} \\
\alpha(\mathrm{k}) &= \frac{(\mathrm{g}(\mathrm{k}+1))^{\mathrm{T}}(\mathrm{g}(\mathrm{k}+1)-\mathrm{g}(\mathrm{k}))}{(\mathrm{g}(\mathrm{k}))^{\mathrm{T}}\mathrm{g}(\mathrm{k})} \\
\alpha(\mathrm{k}) &= \frac{(\mathrm{g}(\mathrm{k})-\mathrm{g}(\mathrm{k}+1))^{\mathrm{T}}\mathrm{g}(\mathrm{k})}{(\mathrm{w}(\mathrm{k})-\mathrm{w}(\mathrm{k}-1))^{\mathrm{T}}(\mathrm{g}(\mathrm{k})-\mathrm{g}(\mathrm{k}-1))}
\end{aligned} \tag{4.11}$$

4.5 Implementation in MATLAB

Consider the example in 0. You can also consider the example of nonlinear function estimation in 0. In this case, instead of the gradient descent and RLS methods, you can use the LM (Levenberg–Marquardt) and CG scripts, which will be discussed further below.

Levenberg–Marquardt script for regulating MLP neural network parameters

```
function  [w1,w2]=LM(e,u,o1,net1,net2,w1,w2,landa)


%% Activation functions
%f1 =@(x)( 8-exp(-x))./(1+exp(-x));
%f2 =@(x)( 8-exp(-x))./(1+exp(-x));
%% Derivative of Activation functions
df1 =@(x)2*exp(-x)./(1+exp(-x)).^2;
%df2 =@(x)2*exp(-x)./(1+exp(-x)).^2;
df2 =@(x)1*exp(-x)./(1+exp(-x)).^2;
delta2=df2(net2);
J2=(o1*delta2');

delta22=e.*delta2;
J22=(o1*delta22');


delta1=(w2(2:end,:)*delta2).*df1(net1);
J1=(u*(delta1)');

delta12=(w2(2:end,:)*delta22).*df1(net1);
J12=(u*(delta12)');

J2=reshape(J2,numel(J2),1);
J22=reshape(J22,numel(J22),1);
J1=reshape(J1,numel(J1),1);
J12=reshape(J12,numel(J12),1);

dw2=(landa*eye(numel(w2))+J2*J2')\J22;%+landa*eye(nume
l(w2))
dw1=(landa*eye(numel(w1))+J1*J1')\J12;

w1=w1+reshape(dw1,size(w1,1),size(w1,2));
w2=w2+reshape(dw2,size(w2,1),size(w2,2));
```

CG snippet for regulating neural network parameters

```
function [w1,w2,g1_old,g2_old,dw1_old,dw2_old]=...

CGD_MLP(eta,e,u,o1,net1,net2,w1,w2,g1_old,g2_old,dw1_o
ld,dw2_old)

%% Activation functions
%f1 =@(x)( 8-exp(-x))./(1+exp(-x));
%f2 =@(x)( 8-exp(-x))./(1+exp(-x));
%% Derivative of Activation functions
df1 =@(x)2*exp(-x)./(1+exp(-x)).^2;
%df2 =@(x)2*exp(-x)./(1+exp(-x)).^2;
df2 =@(x)1*exp(-x)./(1+exp(-x)).^2;
%% w2
delta2=e.*df2(net2);
g2=eta*o1*delta2';
alfa2(:,1)=diag(g2'*(g2-
g2_old))./(diag(g2_old'*g2_old)+0.001);
dw2=g2+dw2_old.*(ones(size(w2,1),1)*alfa2');
w2=w2+dw2;

%% w1
delta1=(w2(2:end,:)*delta2).*df1(net1);
g1=eta*u*(delta1)';
alfa1(:,1)=diag(g1'*(g 8-
g1_old))./(diag(g1_old'*g1_old)+1);
dw1=g1+dw1_old.*(ones(size(w1,1),1)*alfa1');
w1=w1+dw1;
%%
g1_old=g1;
g2_old=g2;
dw1_old=dw1;
dw2_old=dw2;
```

The remaining scripts are similar to those described in the previous chapters. All that is required is a change in the optimization method script. The following script is used, for instance, in the classification problem:

Main scripts in the classification problem

```
%clear all
clc

%%  Load data

load data_classification

%% Input data

Train_data=reshape(Xtrain,size(Xtrain,1),size(Xtrain,2
)*size(Xtrain,3));
Target_train=repmat([1:10],size(Xtrain,2),1);
Target_train=reshape(Target_train,numel(Target_train),
1);

%----------------------------
Test_data=reshape(Xtest,size(Xtest,1),size(Xtest,2)*si
ze(Xtest,3));
Target_test=repmat([1:10],size(Xtest,2),1);
Target_test=reshape(Target_test,numel(Target_test),1);


%% NN Initial parameters
num_neuron=5;
w1=rand(24,num_neuron);
w2=rand(num_neuron+1,10);
dw1_old=zeros(size(w1));
dw2_old=zeros(size(w2));
g1_old=ones(size(w1));
g2_old=ones(size(w2));
eta=0.09;
b1=1;
b2=1;
O_target=zeros(10,1);
rmse_min=100;
landa=1;
%% Train
==================================================
for epoch=1:40
```

```
    %% Train
     E=0;
    nn=randperm(size(Train_data,2));

    for ii=nn
        x=Train_data(:,ii);

[o1,o2,net1,net2]=FeedForward_NN(x,b1,b2,w1,w2);
        O_target=zeros(10,1);
        O_target(Target_train(ii))=1;
        e=O_target-o2;
       %
[w1,w2]=LM(e,[b1;x],[b2;o1],net1,net2,w1,w2,1);
        %w2=LM2(e,[b2;o1],net2,w2,landa);

%[w1,foo]=GD_MLP(e,0.01,[b1;x],[b2;o1],net1,net2,w1,w2
);

        [w1,w2,g1_old,g2_old,dw1_old,dw2_old]=...

CGD_MLP(eta,e,[b1;x],[b2;o1],net1,net2,w1,w2,g1_old,g2
_old,dw1_old,dw2_old);

   E=E+norm(e)^2;
    end
   epoch
  rmse_train=E/size(Train_data,2)
%   MSE_Train(epoch)= E/size(Train_data,2);

   %% Test
     E=0;
    for ii=1:size(Test_data,2)
      x=Test_data(:,ii);
      [o1,o2,net1,net2]=FeedForward_NN(x,b1,b2,w1,w2);
      O_target=zeros(10,1);
```

```
        O_target(Target_test(ii))=1;
        e=O_target-o2;
        E=E+norm(e)^2;
    end
    rmse_test=E/size(Test_data,2)
%    %MSE_Test(epoch)=E/size(Test_data,2);

if rmse_min>rmse_test
   rmse_min=rmse_test;
   w1_best=w1;
   w2_best=w2;
end
end
%%
====================================================

C=zeros(10,10);
for ii=1:size(Test_data,2)
      x=Test_data(:,ii);

[o1,o2,net1,net2]=FeedForward_NN(x,b1,b2,w1_best,w2_be
st);
      O_target=zeros(10,1);
      O_target(Target_test(ii))=1;
      [a,b]=max(abs(o2));
      C(b,Target_test(ii))=C(b,Target_test(ii))+1;
end
C
sum(sum(C.*eye(10)))*100/size(Test_data,2)
```

References

1. M.I. Lourakis, A brief description of the Levenberg-Marquardt algorithm implemented by levmar. Found. Res. Technol. **4** (2005)
2. M.F. Møller, A scaled conjugate gradient algorithm for fast supervised learning. Neural Netw. **6**, 525–533 (1993)

5 Neural Networks Training Based on Genetic Algorithm

5.1 Introduction[1]

5.1.1 What is the Genetic Algorithm (GA)?

The genetic algorithm (GA) is a computer science search technique for estimating solutions and solving problems. A genetic algorithm is a subset of an evolution algorithm that employs biological techniques such as inheritance and mutation.

John Holland invented the genetic algorithm, which is a random optimization (RO) method, in 1967. Later on, this method gained significance as a result of Goldberg's efforts, and it is now considered a well-known technique due to its capabilities.

Genetic algorithms are typically implemented in a computer simulator in which an abstract example of the population (chromosomes) among candidate optimization problem solutions leads to a better solution. Solutions traditionally used to be implemented as series of 0 and 1; however, they are now implemented in other ways. The hypothesis begins with a completely random unique population and progresses through generations. The capacity of the entire population is evaluated in each generation, and some unique subjects are selected at random from the current generation (based on qualifications), are corrected for forming the new generation (deducted or recombined), and then transformed into the current generation in the subsequent iteration of the algorithm.

[1] http://www.beytoote.com/scientific/midanid/genetic-algorithms.html.

A. Mohammadzadeh et al., *Neural Networks and Learning Algorithms in MATLAB*,
Synthesis Lectures on Intelligent Technologies,
https://doi.org/10.1007/978-3-031-14571-1_5

For example, in order to simplify the fluctuations in petroleum prices, external factors and regression values are used, resulting in the following formula: Petroleum price at time t = coefficient 1 of the interest rate at time t + coefficient 2 of the unemployment rate at time t + constant 1. Then, a criterion for determining the best set of coefficients and constants for modeling the petroleum price is applied. This method has two main advantages. First, it is linear, and second, rather than searching through "parameter space," the parameters that are used are determined.

A super formula or plan that expresses something like "petroleum price at t that is a function of a maximum of four variables" is regulated through a genetic algorithm. Then, data for a group of different variables, approximately 20 variables, are provided. The genetic algorithm is then executed to find the best function and variables. The method by which the genetic algorithm works is deceptively simple, very understandable, and, importantly, is a method by which animals are thought to have evolved. Any formula that adheres to the preceding plan is considered a member of the population of possible formulas.

The variables that determine each formula are represented as numbers that are the same as an individual's DNA.

The genetic algorithm engine generates the formula's primary population. Each individual is tested against a set of data, and the most appropriate ones are kept (perhaps 10% of the most appropriate), whereas the others are excluded. The fittest individuals match (move DNA elements) and change (randomly changing DNA elements). The genetic algorithm is observed to be led toward generating formulas that are more precise over time and across many generations. However, neural networks are both nonlinear and nonparametric; the appeal of genetic algorithms is that the final results are more noticeable. The final formula is more observable to a human user, and conventional statistical techniques can be applied to these formulas to present a confidence level of the results. Genetic algorithms are constantly experiencing technological developments, for example, by presenting virus equations that are generated along formulas to violate weak formulas and, as a result, make the population stronger.

The genetic algorithm (GA) is a programming technique that employs genetic evolution as a pattern for problem solving. The input is the problem to be solved, and the solutions are coded based on a pattern known as the fitness function, and each solution evaluates the candidates, the majority of which are selected at random.

The GA is a computer science search technique for finding optimal solutions and solving problems. Genetic algorithms are evolution algorithms that are inspired by biological sciences such as heredity, mutation, accidental selection, natural selection, and hybridity.

Solutions are typically displayed as 0 and 1 binaries; however, there are other ways to display them. Evolution begins with a completely random set of existences and continues in successive generations. The fittest, rather than the best, are selected in each generation.

A list of parameters known as chromosomes or genomes represents a solution to the problem under consideration. Although chromosomes are typically displayed as a string of data, other data structures can also be used. To begin, some characteristics are generated at random for the formation of the first generation. Each characteristic is evaluated in each generation, and fitness is measured through the fitness function.

The next step is to create the community's second generation, which will be based on selection processes and production based on the selected characteristics with genetic operators: chromosomes connect to one another and changes occur.

A pair of parents is selected for each individual. The selections are made in such a way that the fittest are selected, and even the weakest have a chance to be selected so that approaching a local solution is avoided. There are some patterns of selection: roulette wheel selection, tournament selection, etc.

Genetic algorithms typically have a connection probability number ranging from 0.6 to 1 that indicates the likelihood of having a child. With this possibility, organisms are reintegrated. When two chromosomes are connected, a child is produced who is passed down to the next generation. These are carried out in order to identify a qualified candidate for a solution in the next generation. The new children must then be changed. Genetic algorithms typically have a small and fixed change probability of 0.01. Children's chromosomes change or mutate at random based on this probability, particularly by mutating bits in the data chromosome structure.

This process results in a new generation of chromosomes that differ from the previous one. The entire process is repeated for the next generation, couples are selected for integration, and then a third generation is produced. This cycle is repeated until the last stage is reached.

5.1.2 Operators of a Genetic Algorithm

Before a genetic algorithm is employed to solve a problem, two elements must be present: The first is a method for presenting the solution so that the genetic algorithm can operate on it. A solution is traditionally represented as a string of bits, numbers, or characters. The second is a method that uses fitness functions to calculate the quality of each proposed solution. For example, if the problem considers any possible weights fit for the backpack without tearing the backpack apart (see backpack problem), a method for presenting a solution can be thought of as a string of 0 and 1 bits, with 0 and 1 indicating whether or not weight is added to the backpack. The total weight for the proposed solution is used to determine the fitness of the solution.

The genetic algorithm optimization process is based on a random-guided process. This method is based on Darwin's Theory of Gradual Evolution and fundamental ideas. In this method, a collection of target parameters is generated at random for a constant number, called the population. After the simulator is executed, a number representing the standard deviation or fit of that set of data is assigned to that specific member of the population. This is repeated for each and every generated member. Then, by calling genetic algorithm operators such as fertilization, mutation, and selection, the next generation is formed. This continues until convergence criteria is satisfied [1].

There are three criteria commonly known as stopping criteria: (1) The algorithm run-time; (2) The number of generations generated; and (3) The error criterion convergence.

5.1.3 Applications of Genetic Algorithm

- Hydrological routing of runoff flow in dry network
- Assistance in solving multi-criteria decision-making problems
- Multiobjective optimization in water resources management
- Optimization and load arrangement of power distribution networks.

The conditions for terminating genetic algorithms are as follows:

- When a constant number of generations is reached.
- The allocated budget is terminated (computing time/budget).
- An individual (produced child) who meets the bare minimum (least) criteria is discovered.
- The highest degree of fit for children is achieved, or no other better results are obtained.
- Manual checking.

5.2 Genetic Algorithm in MATLAB

- Genetic algorithm calling

To use this algorithm in the command line, enter the following command.

```
[x fval] = ga(@fitnessfun, nvars, options)
```

where

@fitnessfun	is the intended cost function. In other words, the x variables must be identified in order for this function to be minimized.
nvars	is the number of independent variables in the function.

In the results:

fval*	is the final value of the function.
*x	is the point where the function is optimal.

- Using the Genetic Algorithm Toolbox.

This toolbox is a graphical toolbox that allows the user to use this algorithm without having to type anything into a command line. Type the following command and press Enter to access this toolbox, or use the Optimization Toolbox in updated versions (Fig. 5.1).

≫gatool

Fitness function: In this field, enter the function to be optimized.

- Number of variables: The function's number of independent variables.

To begin solving, click the Start in the "run solver".

"Current generation" denotes the number of generations. To stop solving, press "pause."

*The following information is presented in "status & result".

- The function's final value when the algorithm is terminated.
- The reason for the algorithm's termination.
- The point where the function is optimal (Fig. 5.2).

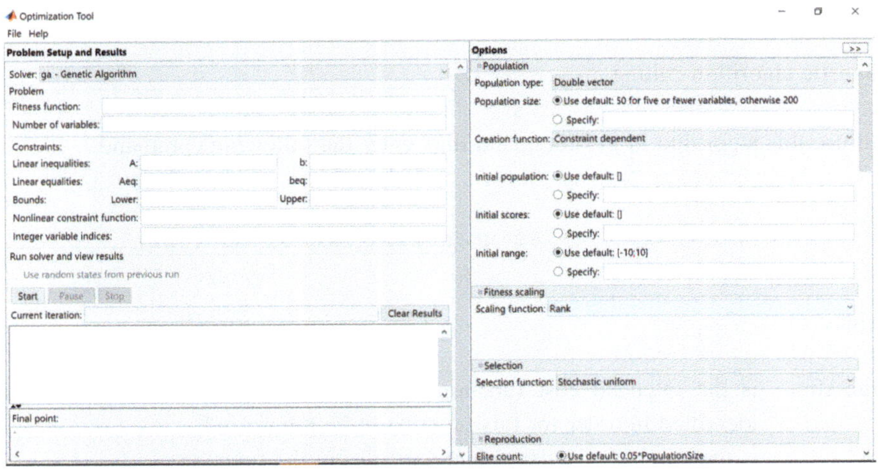

Fig. 5.1 Genetic algorithm graphic toolbox in MATLAB

Fig. 5.2 Results of genetic algorithm running

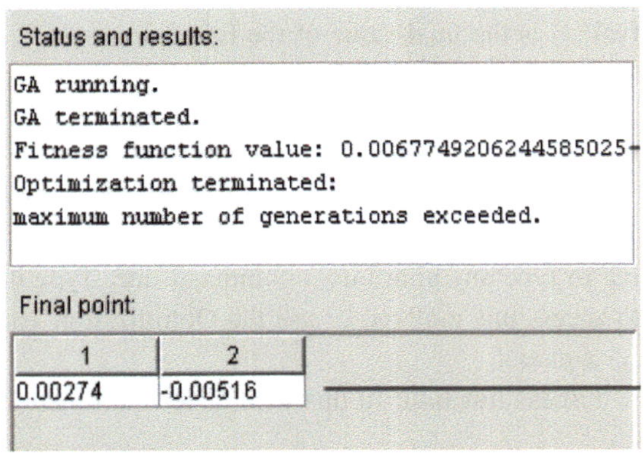

During runtime, you can see various types of information in "plots." You can make appropriate changes and achieve a better result after seeing the results. Choose "best fitness" to see the generation's best and average intended functions (Figs. 5.3 and 5.4).

Example

Consider Rastrigin's function with two independent variables

$$Ras(x) = 20 + x_1^2 + x_2^2 - 10(\cos 2\pi x_1 + \cos 2\pi x_2)$$

This function has a number of local minima, but it also has an absolute minimum at [o.o], as indicated by the vertical line, where the function's value is zero (Fig. 5.5).

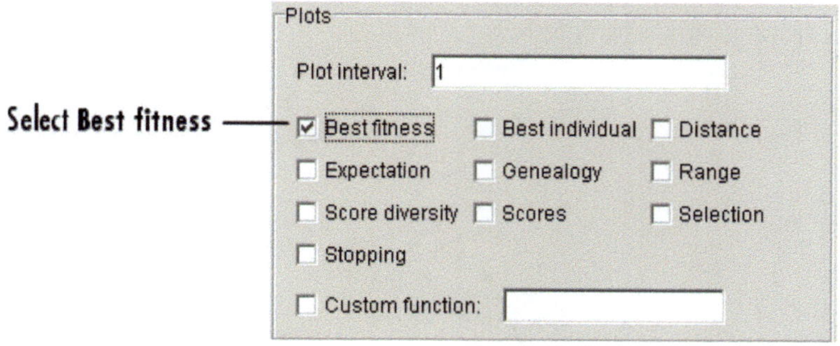

Fig. 5.3 Diagrams that can be drawn while genetic algorithms are running

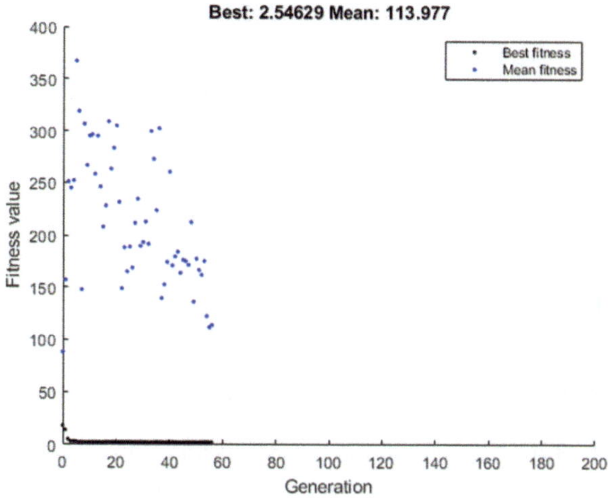

Fig. 5.4 A diagram displaying how the average cost function for the entire population, as well as the cost function for the best fitness, changes

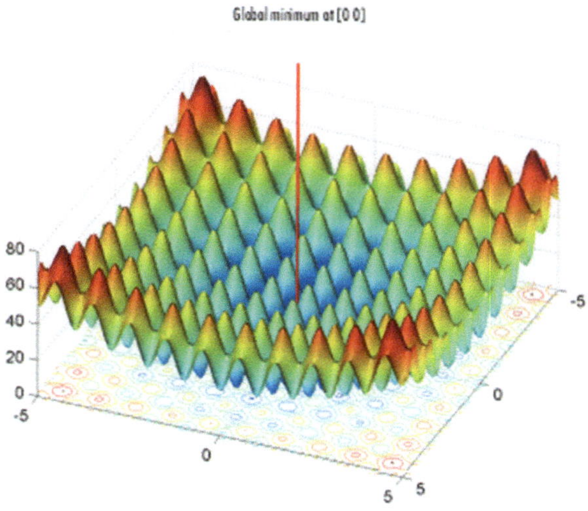

Fig. 5.5 Diagram of Rastrigin's function that has many local minima

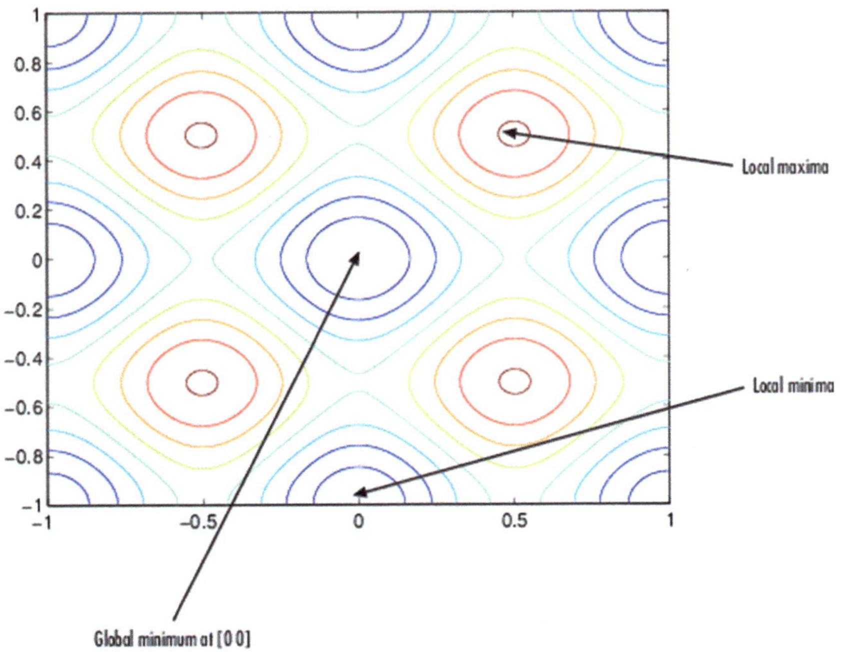

Fig. 5.6 Contours of Rastrigin's function

- Finding minimum point of Rastrigin's function

1. Type "gatoal" in the command line (Fig. 5.6).
   - In "fitness function" enter "@ .rastriginsfcn
   - Enter 2 in "number variable"
   - By pressing "start," the algorithm begins to run.

To find the minimum, use the command line and perform the steps listed below.

```
[x fval reason] = ga(@rastriginsfcn, 2)
```

This returns

```
[x fval reason] = ga(@rastriginsfcn, 2)

x =

    0.0027   -0.0052


fval =

    0.0068


reason =

Optimization terminated:
maximum number of generations exceeded.
```

Key Terms of Genetic Algorithm Toolbox in MATLAB

Individual

For example, for the following function, the vector (2, 3, 1) represents an individual, and f (2, 3, 1) = 51 represents the individual's score. The individual could be recognized as a gene.

- Population and Generation

The population is an array of individuals. For instance, if the function has a variable and the population size is 100, there is a 3×100 array.

- Diversity

The average distance between individuals is defined, and the population with the greatest average distance has a high diversity (Fig. 5.7).

- Fitness Values and Best Fitness Values

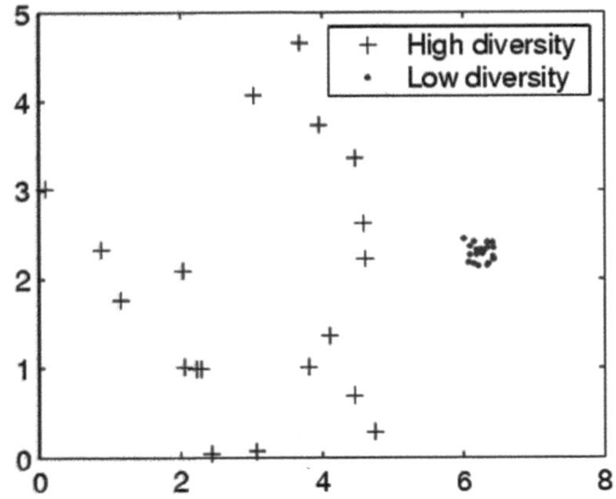

Fig. 5.7 Population diversity diagram

The fitness value of the function varies depending on the individual. The reason for this is that this toolbox is tasked with determining the lowest fitness of the function. The best fitness for a population is the lowest fitness.

- Parents and Children

A genetic algorithm selects some individuals from the current population, parents, and uses them to create the next generation, children, in order to create the next generation. The algorithm selects parents who are fitter.

The minimizing process of the algorithm is explained further below.

1. The algorithm selects the primary population at random.
2. Based on the current generation, the algorithm creates the next generation. To this end, the algorithm runs the stages listed below.
   a. Calculating fitness for each member of the population and assigning a score to each member of the population
   b. Scaling the scores given in order to make better use of the scores and population
   c. Selecting children using their parents. Reproduction through one parent's organ change—genetic mutation or parental combination for reproduction
   d. Selecting parents based on fitness
   e. Replacing the old generation with the new generation.
3. When the stop criteria are met, this algorithm stalls.

- Algorithm's stalling conditions

  To stall the algorithm, five conditions are used.
  - Generations: Once the desired number of generations is obtained (Fig. 5.8).
  - Time limit: When the script running time (in second) reaches the desired number.

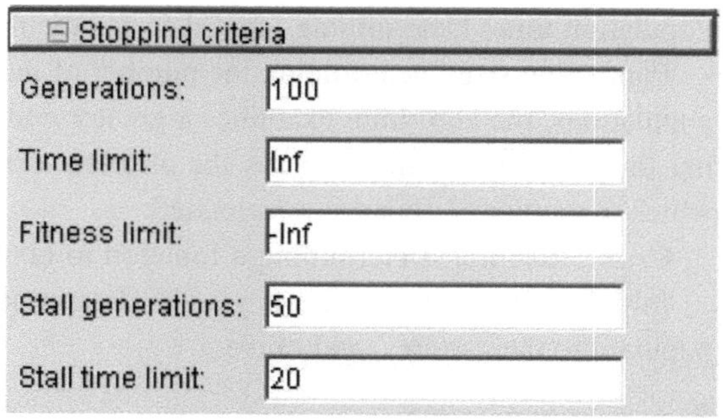

Fig. 5.8 Algorithm's stalling conditions

- Fitness limit: When the function's fitness value at the best point of the current generation falls short of the desired number.

Stall generations: When a better variable is not created in the generations.

Stall time limit: When no better variable is generated in the subsequent generations, the algorithm stalls between two times (in seconds).

The algorithm stalls if any of the following conditions is met first.

Plat option

You can see the work process while running the snippet if you enable the options in "plat."

In "plat," you can use the following functions:

Plot interval (PlotInterval): The number of generations utilized in the "plat."

Best fitness (@gaplotbestf): Plotting the best function value in each generation.

Expectation (@gaplotexpectation): Plotting the expected number of children against the scores of each generation.

Score diversity (@gaplotscorediversity): Plotting a histogram of the scores at each generation.

Stopping (@plotstopping): Plotting stall criteria at each level.

Best individual (@gaplotbestindiv): Plotting the vector of each individual with the best fitness.

Genealogy (@gaplotgenealogy): Plotting the type of reproduction of the next generation.

Scores (@gaplotscores): Plotting each individual's score in their generation.

Distance (@gaplotdistance): Plotting the average distance between individuals in each generation.

Range (@gaplotrange): Plotting the minimum and maximum average value of the function at each generation.

Selection (@gaplotselection): Plotting a histogram of the parents.

Population option

Population time: Determining input data for the fitness function.

Population size: Determining the number of individuals at each generation. With large populations, the algorithm examines a greater number of fitness, and the chances of finding the relative extremum versus the absolute extremum are reduced, but the pace with which a solution is found is accelerated.

Create function: Determining a function to obtain the initial population.

Initial population: It allows you to determine the number of people in the initial population, their scores, and range.

Fitness scaling option

This function converts each fitness's score to a specific scale suitable for the selection function.

In this section, the default is "rank." Instead of using scores, scaling is based on each individual's layout.

Proportional: Scaling based on each individual's score.

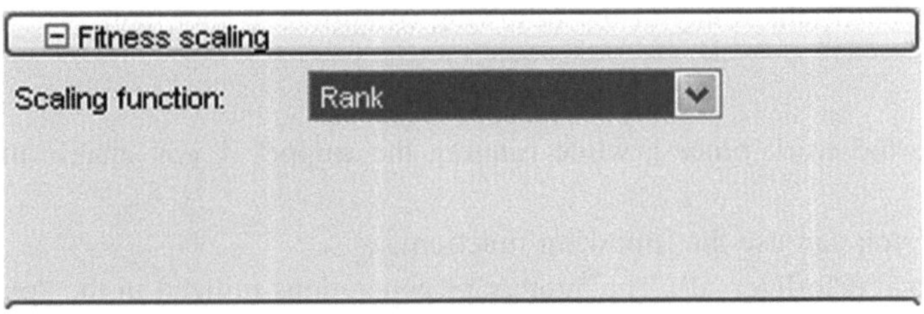

Selection-option

This option specifies how the algorithm selects parents for the next generation.

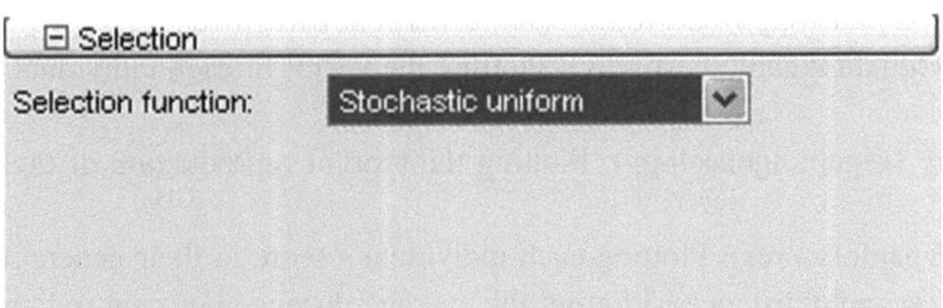

Reproduction

This option specifies how the algorithm selects children for the next generation.

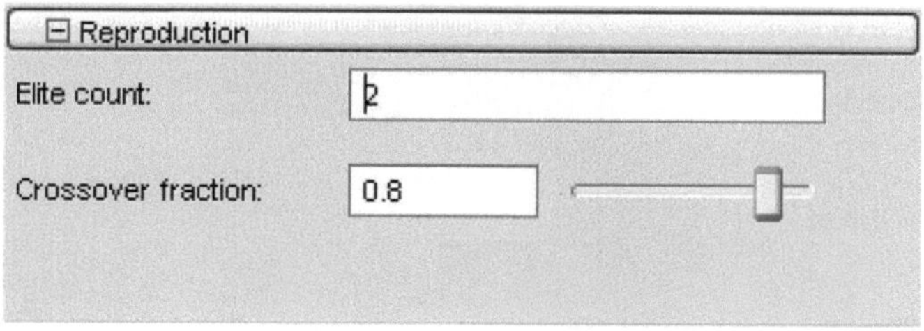

Mutation option

This option determines how the algorithm makes a small random change in each individual in the event of a genetic mutation. Genetic mutation enables the algorithm to utilize a larger space.

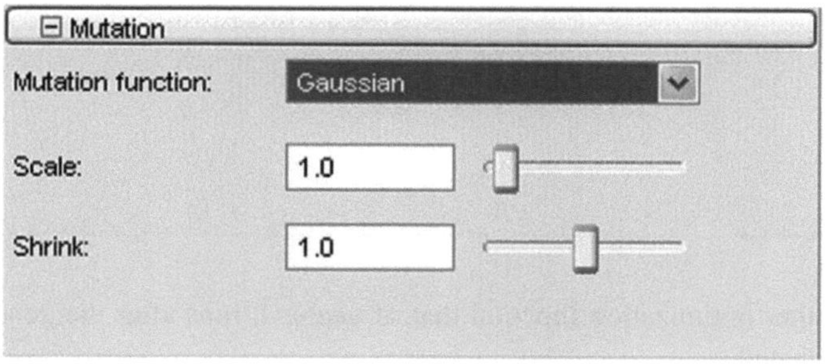

Crossover

This option determines how the algorithm creates a new individual out of two individuals.

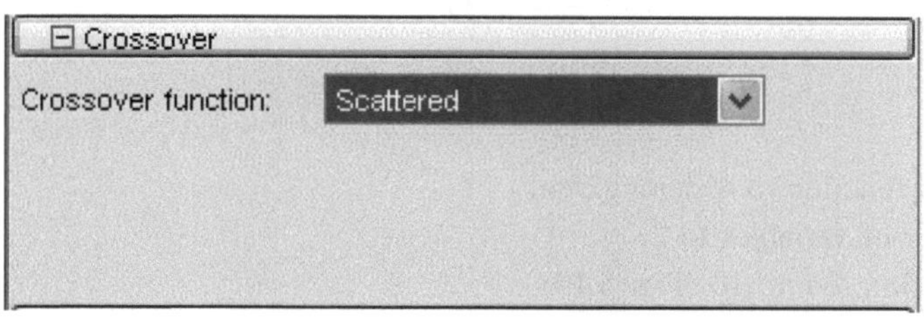

Migration option

This option determines how individuals migrate across subpopulations. When the population size exceeds one, migration occurs. When this occurs, the fittest individuals in one population are replaced by the least fit in another. This substitute is done by copying.

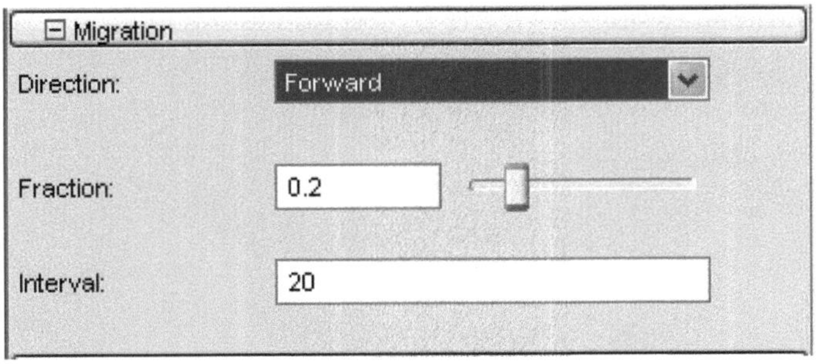

Hybrid function option

This is another optimization function that, if enabled, runs after the genetic algorithm has been terminated.

Example

In this case, the Rosenbrock function's genetic algorithm finds a point near the minimum (Fig. 5.9).

$$f(x, y) = (a - x)^2 + b\left(y - x^2\right)^2$$

Set Fitness function to @dejong2fcn.
Set Number of variables to 2.
Set Population size to 10 (Fig. 5.10).

"Hybrid Function" can then be used to improve this optimization (Fig. 5.11).
And, find a solution that is closer to the minimum point.

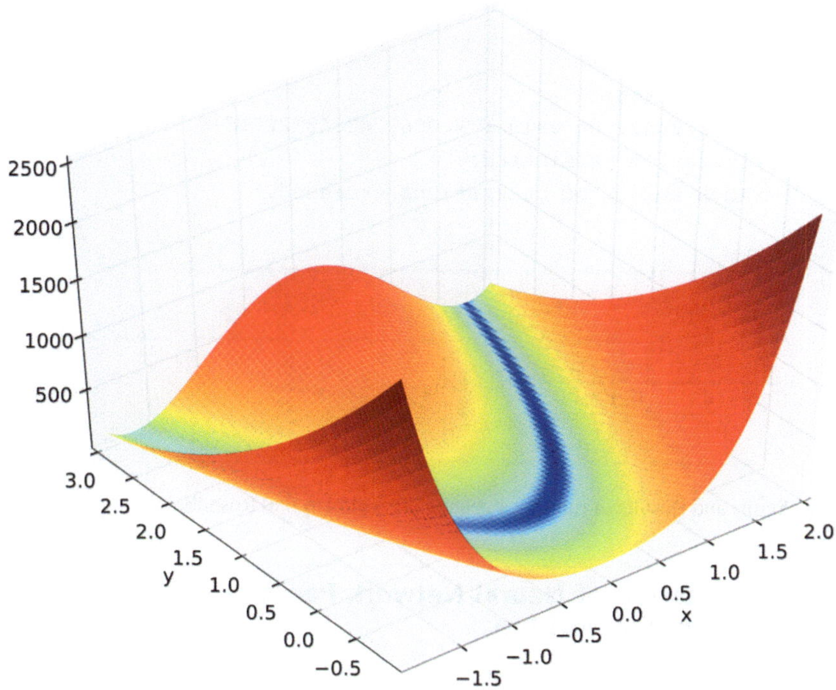

Fig. 5.9 Diagram of Rosenbrock function

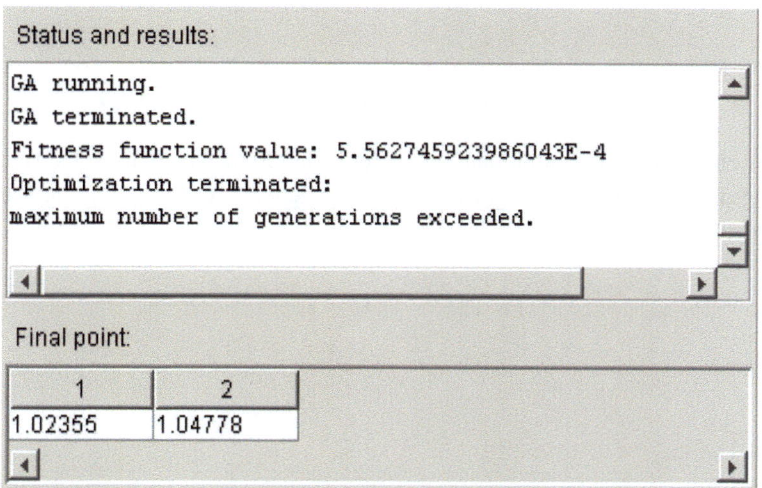

Fig. 5.10 Results of running a genetic algorithm to find Rosenbrock function's minimum

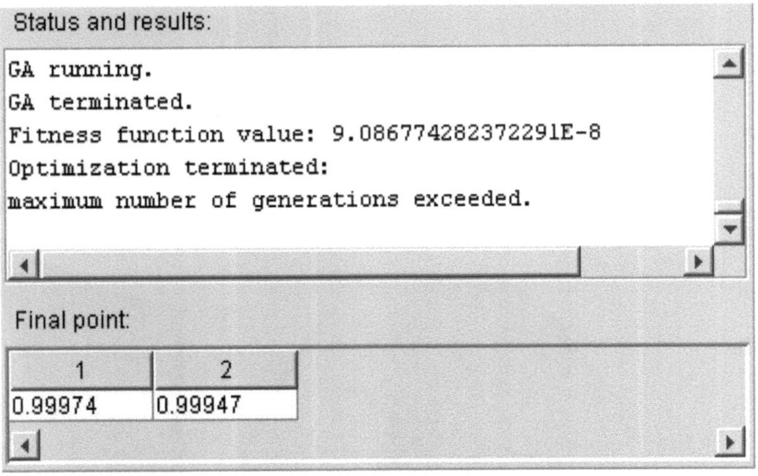

Fig. 5.11 Status and Results of running a genetic algorithm to find Rosenbrock function's minimum

5.3 Optimization of Neural Network Parameters Based on Genetic Algorithm

Using MLP neural networks, it is intended to estimate the following nonlinear function:

$$y(t) = y(t-1)/(1 + y(t-1)^2) + u(t-1)^3 \tag{5.1}$$

where

$$\mathrm{u} = \sin(2\pi \mathrm{n}/100), \quad \mathrm{ts} = 100/500, \quad \mathrm{n} = 0 : \mathrm{ts} : 100$$

To this end, 500 input-output data are generated, 400 of which are selected at random as training data and the remaining as test data. To generate test and training data, run the following snippet.

Initialization snippet and data generation for testing and training

```
clear all
clc
global Test_data Target_test Train_data Target_train
num_out=1;
num_in=2;
%% ========================================
ts=100/500;
n=0:ts:100;
u(:,1)=sin(2.*pi.*n./100);
y(2,1)=0;
for t=3:length(u);
 y(t,1) =y(t-1,1)/( 1 + y(t-1,1)^2) + u(t-1)^3;
end
%% Input data
data=[[u(1:end)]  [0;y(1:end-1)]];
nn=randperm(500);
Train_data=data(nn(1:400),:);
Target_train=y(nn(1:400));

Test_data=data(nn(401:500),:);
Target_test=y(nn(401:500));
%% ==========================================
%% NN Initial parameters
n_neroun=10;
numel_w1=10*2;
numel_w2=10;
P0=rand(100,numel_w1+numel_w2);
```

After running "initialization snippet and data generation for testing and training", you can run the following main script.

The main snippet for regulating neural network parameters based on genetic algorithm

```
clc
global Test_data Target_test
%% Train
   num_v=size(P0,2);
    options = gaoptimset('InitialPopulation',P0,...
        'StallGenLimit',100,'PopulationSize',100,...

'Generations',100,'PopulationType','doubleVector',...


    'CrossoverFcn',@myfun,'PlotFcns',{ @gaplotstopping,
    @gaplotbestf } );
        [w,Fval,exitFlag,Output]=ga(@(w)
    FitFcn2(w),num_v,...
                [],[],[],[],[],[],[],[],options);

%% plot
fprintf('The number of generations was : %d\n',
Output.generations);
fprintf('The number of function evaluations was :
%d\n', Output.funccount);
fprintf('The best function value found was : %g\n',
Fval);

%% Test
w1=reshpe(w(1:2*n_neroun),2,n_neroun);
w2(:,1)=w(2*n_neroun+1:end);
for ii=1:size(Test_data,1)
  x=Test_data(ii,:)';
  yest(ii)=FeedForward_NN(x,w1,w2);
end
plot(yest,'--r')
hold on
plot(Target_test,'--b')
```

The following functions are used in this snippet:

FeedForward_NN function

```
function [o2]=FeedForward_NN(u,w1,w2)
%% Activation functions
f1 =@(x)( 8-exp(-x))./(1+exp(-x));
%f2 =@(x)( 8-exp(-x))./(1+exp(-x));
f2 =@(x) 1./(1+exp(-x));
%% Hidden Layer
net1=w1'*u;
o1=f1(net1);
%% Out Layer
net2=w2'*o1;
o2= (net2);
```

FitFcn2function

```
function EE=FitFcn2(w)
global   Train_data Target_train
n_neroun=10;
w1=reshape(w(1:2*n_neroun),2,n_neroun);
w2(:,1)=w(2*n_neroun+1:end);
EE=0;
  for ii=1:size(Train_data,1)
    x=Train_data(ii,:)';
    o=FeedForward_NN(x,w1,w2);
    e=Target_train(ii)-o;
    EE=EE+sum(e.^2);
  end
 EE=EE/size(Train_data,1);
```

Myfun function (a function written for a custom crossover).

```
function xoverKids = myfun(parents, options, nvars,
FitnessFcn, ...
    unused,thisPopulation)

R=rand(length(parents)/2,1);
parents=reshape(parents,length(parents)/2,2);
xoverKids=thisPopulation(parents(:,1),:).*(R*ones(1,nv
ars))+...
          (( 8-
R)*ones(1,nvars)).*thisPopulation(parents(:,2),:);

end
```

Figures 5.12 and 5.13 show the genetic algorithm's output diagrams as well as the estimation result. As shown, the neural network's estimated output matches the target output quite well.

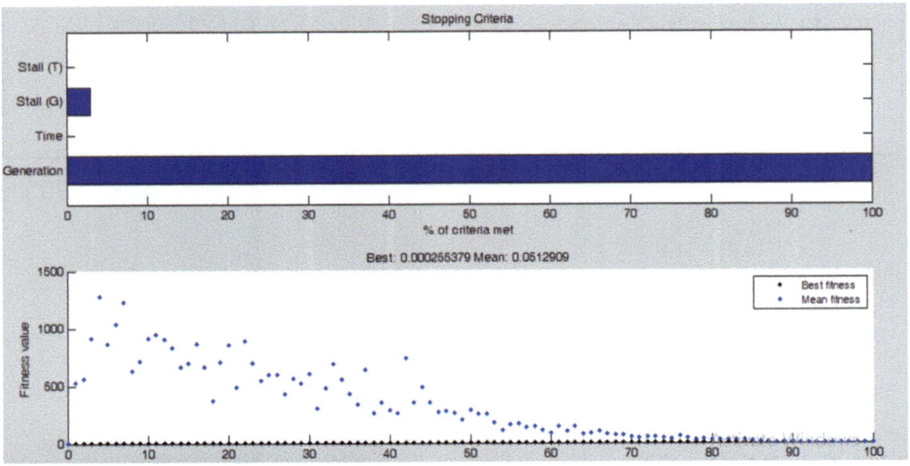

Fig. 5.12 Genetic algorithm output

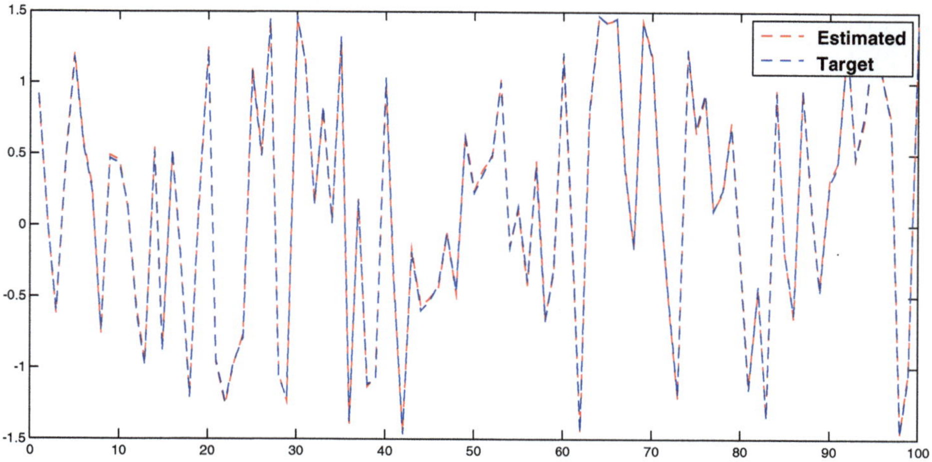

Fig. 5.13 Diagram of target output and estimated output using MLP neural network

Reference

1. H. Adeli, S.-L. Hung, *Machine Learning: Neural Networks, Genetic Algorithms, and Fuzzy Systems* (Wiley, Hoboken, 1994)

Neural Network Training Based Particle Swarm Optimization (PSO)

6

6.1 Introduction

Eberhart and Kennedy first proposed this algorithm in 1995 as an indefinite technique for function optimization. The algorithm was inspired by the motion of flocks of birds looking for food [1, 2] (Fig. 6.1).

A flock of birds searches for food at random in a place. There is only one piece of food there. None of the birds known where the food is. One of the best strategies may be following the bird with the least distance to the food. This strategy is actually an algorithm. Each solution, which is called a particle in this algorithm stands for a bird. Each particle has a certain merit that is measured by a merit function. The closer it gets to the target, food in the case of a flock of birds, the more merit it has in the search space. Furthermore, each particle has a speed that determines how the particle moves. Following the optimal particles in the current state, each particle maintains its position through the problem space. At first, some particles are generated at random, and then generations are updated in an attempt to find the optimal solution. Each particle is updated using two bests in each step. The first is the best position the particle has ever reached. The position is identified and maintained. The other best, which is used by the algorithm as pbest, is the optimal position that is obtained by the population of particles. This position is indicated by gbest [3].

6.2 Algorithm Formulation

Personal Best (within a given neighborhood or the best position that particle has ever achieved) or Global Best can influence the behavior of each particle in the formulation of this algorithm (best particle among particles). In the beginning, particles are randomly

A. Mohammadzadeh et al., *Neural Networks and Learning Algorithms in MATLAB*,
Synthesis Lectures on Intelligent Technologies,
https://doi.org/10.1007/978-3-031-14571-1_6

Fig. 6.1 Particle swarm optimization

initialized throughout the search space. These initial positions are also known as the personal-best position of the particles (pbest). In the next stage, the best particle is chosen among the existing particles and is known as the best solution (best). Then, the particles move across the search space until the termination criteria are met. This movement entails applying a rate equation to the particles, which determines the position of each particle.

The new fitness value generated by the particle is compared to the particle's **pbest** value. If the particle's new position has a higher fitness, it replaces the **pbest** position. The same steps are taken for **gbest**.

Where

X*t*	is the old (current) position.
Xt + 1	is the particle's new (future) position.

If the value of the future velocity is added to the current position, the new (future) position is obtained.

Vt + 1	The random value of three components (old (current) velocity, best, and pbest) is used to calculate future velocity.

Therefore, the mathematical relation of particle movement is as follows [4]:

$$\begin{aligned} \text{Pnew} &= \text{Pold} + \text{Vnew} \\ \text{Vnew} &= \text{Vold} + \text{C1} \times \text{R1} \times (\text{Plocal best-Pold}) \\ &\quad + \text{C2} \times \text{R2} \times (\text{Pglobal best-Pold}) \end{aligned} \tag{6.1}$$

where C1 and C2 are constant and positive values, while R1 and R2 are random numbers that are normally generated in the [0,1] range. To improve search capability, an inertia weight parameter is added as a factor to the algorithm velocity parameter as follows:

$$\text{Vnew} = \text{W} \times \text{Vold} + \text{C1} \times \text{R1} \times (Plocalbest - Pold)$$

$$+ C2 \times R2 \times (Pglobalbest - Pold) \tag{6.2}$$

The effect of the particle velocity in the previous step on the current velocity is determined by the inertia weight. Thus, increasing the inertia weight improves the algorithm's general search capability and allows it to examine more space. With smaller values of inertia weight, however, the space under study is limited, and the search is performed within this limited space.

Therefore, the algorithm typically begins with a larger number of inertia weights, resulting in a large space search at the start of the algorithm, and this weight decreases over time, focusing the search on small space in the final steps.

Pseudocode of particle swarm optimization[5]

Algorithm gbest PSO (Initialize)
gbest=X_0
for i=0 **to** $N_{particles}$ **do**
  pbest$_i$=X_i *(initialize randomly)*
  fitness$_i$=f(X_i)
  if fitness$_i$ < f(gbest) **then**
    gbest=X_i
  end if
end for

Algorithm gbest PSO (Main loop)
repeat
  for i=0 to $N_{particles}$ do
    V_i=W*V_i+c_1*r_1*(pbest$_i$-X_i)+c_2*r_2*(gbest-x_i)
    if $V_i \notin V_{acceptable}$ then
      correct V_i
    end if
    X_i=X_i+V_i
    fitness$_i$ = f(X_i)
    if fitness$_i$ < f(pbest$_i$) then
      pbest$_i$=X_i
    end if
    if fitness$_i$ < f(gbest) then
      gbest=X_i
    end if
  end for
untill terminate criteria

6.3 Implementation in MATLAB

Using MLP neural networks, it is intended to estimate the following nonlinear function:

$$y(t) = y(t-1)/(1 + y(t-1)^2) + u(t-1)^3 \tag{6.3}$$

where

$$\mathrm{u} = \sin(2\pi \mathrm{n}/100), \quad \mathrm{ts} = 100/500, \quad \mathrm{n} = 0 : \mathrm{ts} : 100$$

To this end, 500 input-output data are generated, 400 of which are selected at random as training data and the remaining as test data. To generate test and training data, run the following snippet.

Initialization snippet and data generation for testing and training

```
clear all
clc
global Test_data Target_test Train_data Target_train
num_out=1;
num_in=2;
%%
======================================================
===============
ts=100/500;
n=0:ts:100;
u(:,1)=sin(2.*pi.*n./100);
y(2,1)=0;
for t=3:length(u);
 y(t,1) =y(t-1,1)/( 1 + y(t-1,1)^2) + u(t-1)^3;
end

%% Input data===
data=[[u(1:end)]  [0;y(1:end-1)]];
nn=randperm(500);
Train_data=data(nn(1:400),:);
Target_train=y(nn(1:400));

Test_data=data(nn(401:500),:);
Target_test=y(nn(401:500));
%% NN Initial parameters
b1=0;
b2=0;
num_neuron=5;
w1=rand(num_in+1,num_neuron);
w2=rand(num_neuron+1,num_out);
```

After "initialization snippet and data generation for testing and training" is executed, you can run the following main script.

The main script for regulating neural network parameters based on PSO algorithm

```
clc
global Test_data Target_test
num_neuron=5;
num_out=1;
num_in=2;
num_v=(num_neuron+1)+(num_in+1)*num_neuron;
%% Train
pop=rand(100,num_v);
max_it=150;

   [Gbest,e_Gbest]=PSO(pop,max_it);


%% Test
w1=reshape(Gbest(:,1:(num_in+1)*num_neuron),num_in+1
,num_neuron);
w2(:,1)=Gbest((num_in+1)*num_neuron+1:end);
b1=0;
b2=0;
for ii=1:size(Test_data,1)
   x(:,1)=Test_data(ii,:)';
   yest(ii)=NN(x,b1,b2,w1,w2);
end
plot(yest,'--r')
hold on
plot(Target_test,'--b')
```

The following functions are used in this script:

MLP neural network modeling function called NN

```
function o2=NN(u,b1,b2,w1,w2)

%% Activation functions
f1 =@(x)( 8-exp(-x))./(1+exp(-x));
f2 =@(x)( 8-exp(-x))./(1+exp(-x));
%f2 =@(x) 1./(1+exp(-x));
%% Hidden Layer
u1=[b1;u];
net1=w1'*u1;
o1=f1(net1);
%% Out Layer
u2=[b2;o1];
net2=w2'*u2;
o2=(net2);
```

FitFcn cost function

```
function EE=FitFcn(w)
global  Train_data Target_train
num_neuron=5;
num_in=2;
w1=reshape(w(:,1:(num_in+1)*num_neuron),num_in+1,num
_neuron);
w2(:,1)=w((num_in+1)*num_neuron+1:end);
b1=0;
b2=0;
 EE=0;
  for ii=1:size(Train_data,1)
      x(:,1)=Train_data(ii,:)';
      o=NN(x,b1,b2,w1,w2);
      e=Target_train(ii)-o;
      EE=EE+sum(e.^2);
  end
 EE=EE/size(Train_data,1);
```

PSO function

```
function   [Gbest,e_Gbest]=PSO(pop,max_it)


%%
best_EE=0.0001;
w=0.7;
c1=2;
c2=2;
e_Gbest=inf;
e_Pbest=inf*ones(size(pop,1),1);
v=zeros(size(pop));
for epoch=1:max_it

%% FitFcn
        for ii=1:size(pop,1)
            EE(ii)=FitFcn(pop(ii,:));

            %---------- Pbest-------------

             if EE(ii)<e_Pbest(ii)
                Pbest(ii,:)=pop(ii,:);
                e_Pbest(ii)=EE(ii);
             end
        end
%% ---------- Gbest-------------
 [a b]=min(EE);
if a<e_Gbest
    Gbest=pop(b,:);
    e_Gbest=a;
end
e_Gbest
%%
v=w*(v+c1*rand*(Pbest-
pop)+c2*rand*(repmat(Gbest,size(pop,1),1)-pop));
pop=pop+v;

if e_Gbest<best_EE
    break
end
%%
end % epoch
```

Figure 6.2 depicts target output diagrams as well as estimated output. As shown in Fig. 6.3, the final RMSE value for the best PSO solution is very small.

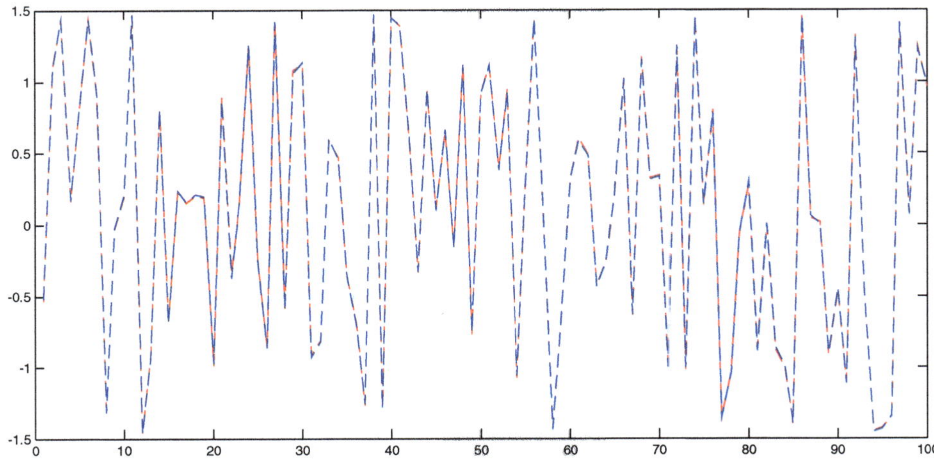

Fig. 6.2 Diagram of target output and estimated output using neural network

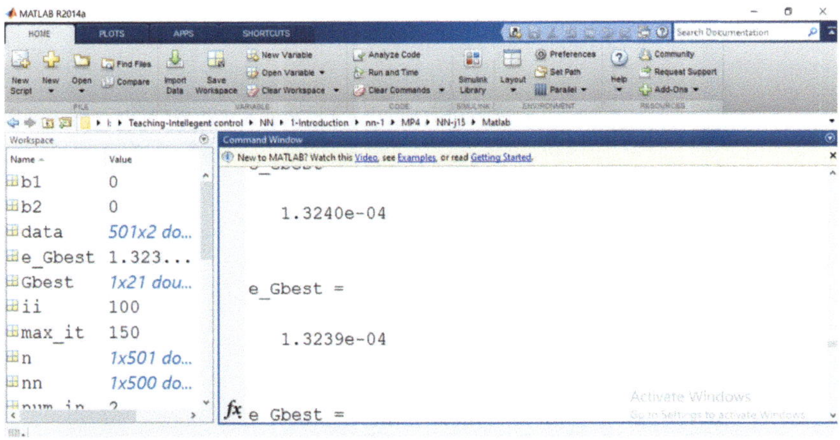

Fig. 6.3 Final optimization result using PSO

References

1. J. Kennedy, Particle swarm optimization, in *Encyclopedia of Machine Learning* (Springer, Berlin, 2011), pp. 760–766
2. Available: http://en.wikipedia.org/wiki/Swarm_intelligence
3. M. Clerc, *Particle Swarm Optimization*, vol. 93 (Wiley, Hoboken, 2010)
4. A. Lazinica, *Particle Swarm Optimization* (InTech Kirchengasse, 2009)
5. K.-L. Du, M. Swamy, Particle swarm optimization, in *Search and Optimization by Metaheuristics* (Springer, Berlin, 2016), pp. 153–173

Neural Network Training Based on UKF 7

7.1 UKF Algorithm

Consider the following nonlinear system:

$$\begin{cases} \dot{x} = f(x, u) + N \\ y = h(x, u) + v \end{cases}, \begin{cases} x_{k+1} = f(x_k, u_k) + N_k \\ y_{k+1} = h(x_k, u_k) + v_k \end{cases} \tag{7.1}$$

where N represents the process noise with variance Q and noise v. Variance R is used for measurement.

The extended Kalman filter is inefficient when the state transition and observations models, i.e., the predict and update functions f and h, are highly nonlinear [1]. This is due to an increase in covariance during the linearization of the nonlinear model. To collect a minimal sample set of points around the mean, the Unscented Kalman filter employs a definite sampling method known as Unscented Transform. These points are then fed into the nonlinear function, which produces a new mean and covariance. The obtained result provides more precise mean value and covariance for definite systems [2]. For posterior statistics, this method is known as the Monte-Carlo method or Taylor series expansion. In fact, this method allows you to avoid directly calculating the Jacobian matrix, which is used for extremely complex functions.

Prediction

The UKF method, like EKF, compares the prediction phase to a linear update that is independent of the UKF update. The mean and process covariance are used to calculate state and covariance estimates [3].

A. Mohammadzadeh et al., *Neural Networks and Learning Algorithms in MATLAB*, Synthesis Lectures on Intelligent Technologies,
https://doi.org/10.1007/978-3-031-14571-1_7

$$\mathbf{x}^a_{k-1|k-1} = [\hat{\mathbf{x}}^{\mathrm{T}}_{k-1|k-1} \quad E[\mathbf{w}_k^{\mathrm{T}}]\ ]^{\mathrm{T}}$$
$$\mathbf{P}^a_{k-1|k-1} = \begin{bmatrix} \mathbf{P}_{k-1|k-1} & 0 \\ 0 & \mathbf{Q}_k \end{bmatrix} \tag{7.2}$$

Using state and covariance from the state after L, a set of 2L + 1 can be obtained.

$$\begin{aligned} \chi^0_{k-1|k-1} &= \mathbf{x}^a_{k-1|k-1} \\ \chi^i_{k-1|k-1} &= \mathbf{x}^a_{k-1|k-1} + \left(\sqrt{(L+\lambda)\mathbf{P}^a_{k-1|k-1}}\right)_i, \quad i = 1, \ldots, L \\ \chi^i_{k-1|k-1} &= \mathbf{x}^a_{k-1|k-1} - \left(\sqrt{(L+\lambda)\mathbf{P}^a_{k-1|k-1}}\right)_{i-L}, \; i = L+1, \ldots, 2L \end{aligned} \tag{7.3}$$

then

$$\left(\sqrt{(L+\lambda)\mathbf{P}^a_{k-1|k-1}}\right)_i \left(\sqrt{(L+\lambda)\mathbf{P}^a_{k-1|k-1}}\right)_i \tag{7.4}$$

The ith column of the matrix is the square root of $\mathbf{P}^a_{k-1|k-1}(L + \lambda)\mathbf{P}^a_{k-1|k-1}$. Considering the definition of the square root, A is obtained in the matrix B:

$$B \triangleq \mathbf{A}\mathbf{A}^{\mathrm{T}} \tag{7.5}$$

Methods such as Cholesky decomposition must be used to calculate the square root numerically.

The obtained points are fed into the transfer function f:

$$\chi^i_{k|k-1} = f(\chi^i_{k-1|k-1}) \quad i = 0, \ldots, 2L \tag{7.6}$$

Therefore, $f : R^L \rightarrow R^{|\mathbf{x}|}$, The state estimate and covariance are calculated using weighted points.

$$\begin{aligned} \hat{\mathbf{x}}_{k|k-1} &= \sum_{i=0}^{2L} W^i_s \chi^i_{k|k-1} \\ \mathbf{P}_{k|k-1} &= \sum_{i=0}^{2L} W^i_c \, [\chi^i_{k|k-1} - \hat{\mathbf{x}}_{k|k-1}][\chi^i_{k|k-1} - \hat{\mathbf{x}}_{k|k-1}]^{\mathrm{T}} \end{aligned} \tag{7.7}$$

The weights for states and covariance are derived from the following relationships:

$$\begin{aligned} W^0_c &= \frac{\lambda}{L+\lambda} + (1 - \alpha^2 + \beta) \\ W^i_s &= W^i_c = \frac{1}{2(L+\lambda)} \\ \lambda &= \alpha^2(L+\kappa) - L \end{aligned} \tag{7.8}$$

The densities of the points are controlled by α and κ. β Represents x distribution. If x distribution is Gaussian, the natural values are $\alpha = 10^{-3}, \kappa = 0$ and $\beta = 2$. $\beta = 2$ is optimized.

Update

Except when the mean and covariance of the observed noise are available, state and covariance predictions are obtained as described above.

$$\begin{aligned} \mathbf{x}^a_{k|k-1} &= \left[\hat{\mathbf{x}}^{\mathrm{T}}_{k|k-1} \quad E[\mathbf{v}^{\mathrm{T}}_k]\right]^{\mathrm{T}} \\ \mathbf{P}^a_{k|k-1} &= \begin{bmatrix} \mathbf{P}_{k|k-1} & 0 \\ 0 & \mathbf{R}_k \end{bmatrix} \end{aligned} \tag{7.9}$$

As mentioned above, consider a set consisting of 2L + 1 points

$$\begin{aligned} \chi^0_{k|k-1} &= \mathbf{x}^a_{k|k-1} \\ \chi^i_{k|k-1} &= \mathbf{x}^a_{k|k-1} + \left(\sqrt{(L+\lambda)\mathbf{P}^a_{k|k-1}}\right)_i, \quad i = 1, \ldots, L \\ \chi^i_{k|k-1} &= \mathbf{x}^a_{k|k-1} - \left(\sqrt{(L+\lambda)\mathbf{P}^a_{k|k-1}}\right)_{i-L}, \quad i = L+1, \ldots, 2L \end{aligned} \tag{7.10}$$

The points can be calculated independently if the UKF predict is used, as shown below.

$$\chi_{k|k-1} := [\chi^{\mathrm{T}}_{k|k-1} \quad E[\mathbf{v}^{\mathrm{T}}_k]\,]^{\mathrm{T}} \pm \sqrt{(L+\lambda)\mathbf{R}^a_k} \tag{7.11}$$

then

$$\mathbf{R}^a_k = \begin{bmatrix} 0 & 0 \\ 0 & \mathbf{R}_k \end{bmatrix} \tag{7.12}$$

These points are given to the h function as input.

$$\gamma^i_k = h(\chi^i_{k|k-1}) \quad i = 0..2L \tag{7.13}$$

The observation and covariance of predict observations are calculated using weighted points.

$$\begin{aligned} \hat{\mathbf{z}}_k &= \sum_{i=0}^{2L} W^i_s \gamma^i_k \\ \mathbf{P}_{z_k z_k} &= \sum_{i=0}^{2L} W^i_c [\gamma^i_k - \hat{\mathbf{z}}_k][\gamma^i_k - \hat{\mathbf{z}}_k]^{\mathrm{T}} \end{aligned} \tag{7.14}$$

The following formula is used to calculate the cross covariance of the states and observations:

$$\mathbf{P}_{x_k z_k} = \sum_{i=0}^{2L} W_c^i \, [\chi_{k|k-1}^i - \hat{\mathbf{x}}_{k|k-1}][\gamma_k^i - \hat{\mathbf{z}}_k]^{\mathrm{T}} \tag{7.15}$$

Which is used to calculate the result of Unscented Kalman Filter (UKF).

$$K_k = \mathbf{P}_{x_k z_k} \mathbf{P}_{z_k z_k}^{-1} \tag{7.16}$$

The updated state, like the Unscented Kalman Filter (UKF), is calculated from the sum of the predicted state and the weighting of the Kalman result.

$$\hat{\mathbf{x}}_{k|k} = \hat{\mathbf{x}}_{k|k-1} + K_k(\mathbf{z}_k - \hat{\mathbf{z}}_k) \tag{7.17}$$

Furthermore, the updated covariance equals the difference between the predicted and calculated covariance, which is weighted by Kalman.

$$\mathbf{P}_{k|k} = \mathbf{P}_{k|k-1} - K_k \mathbf{P}_{z_k z_k} K_k^{\mathrm{T}} \tag{7.18}$$

7.2 Implementation in MATLAB

In order to regulate the MLP neural network parameters based on this algorithm, consider the following example, in which a significant nonlinear function is estimated using the MLP neural network. The following indicates the dynamics of the intended nonlinear function:

$$\begin{aligned} y(k+1) &= f(y(k), y(k-1), y(k-2), u(k), u(k-1)) \\ f(x_1, x_2, x_3, x_4, x_5) &= \frac{x_1 x_2 x_3 x_5 (x_3 - 1) + x_4}{1 + x_2^2 + x_3^2} \end{aligned} \tag{7.19}$$

The 1000-point input is considered at random in the range $[-1,\ 1]$.

When regulating the neural network parameters, the output of the neural network is considered as the space in Fig. 7.1 [4]:

$$\begin{cases} w_{k+1} = w_k + N_k \\ y_{k+1} = h(w_k, u_k) + v_k \end{cases} \tag{7.20}$$

where w_k is the neural network's weight at time k. u_k is the neural network input, and $h(w_k, u_k)$ is a nonlinear function based on the neural network structure. In other words, y_{k+1} is the neural network's output for u_k input and w_k weights.

The MATLAB snippets are listed below. Run the main MATLAB script. It is important to note that the rest of the functions must be in the same folder as this snippet.

Main MATLAB script

```
clear all
clc

%%  Generate Training and Test data
u(:,1)= 8-2*rand(1000,1);
y(3,1)=0;
% h(x1,x2,x3,x4,x5)=( x1*x2*x3*x5(x3-1)+x4 ) /
(1+x2^2+x3^2)
for t=4:length(u);

  y(t) =(y(t-1)*y(t-2)*y(t-3)*u(t-2)*(y(t-3)-1)+u(t-
1))/(1 + y(t-2)^2+y(t-3)^2);
end

%% Input data
data=[[0;y(1:end-1)] [0;0;y(1:end-2)] [0;0;0;y(1:end-
3)] [0;u(1:end-1)] [0;0;u(1:end-2)] ];

nn=randperm(1000);
Train_data=data(nn(1:700),:);
Target_train=y(nn(1:700));

Test_data=data(nn(701:end),:);
Target_test=y(nn(701:end));

%% NN Initial parameters
num_neuron=5;
num_in=5;
num_v=num_neuron+num_neuron*num_in;
w=rand(num_v,1);
P=eye(num_v);
q=0.01;         %std of process
r=0.01;         %std of measurement
Q=q^2*eye(num_v); % covariance of process
R=r^2;          % covariance of measurement
%%
f1 =@(x)( 8-exp(-x))./(1+exp(-x));
f2 =@(x)( 8-exp(-x))./(1+exp(-x));
h=@(x,uu)
f2(x(1:num_neuron)'*f1(reshape(x(num_neuron+1:end),num
_in,num_neuron)'*uu));
f=@(x,uu) x;
min_mse=100;
```

```
%% Train
====================================================
for epoch=1:10
    epoch
    %% Train
    E=0;
    nn=randperm(700);
    kk=1;
    for ii=nn
        x=Train_data(ii,:)';
        [w, P] = ukf(f,w,P,h,Target_train(ii),Q,R,x);
        o2=h(w,x);
        e=Target_train(ii)-o2;
        E=E+e^2;
    end
  MSE_Train= E/700
  % MSE_Train(epoch)= E/700;

   %% Test
     E=0;
    for ii=1:300
      x=Test_data(ii,:)';

      o2=h(w,x);
      e=Target_test(ii)-o2;
      E=E+e^2;

    end
    MSE_Test= E/300
   %%
   if MSE_Test<min_mse
       w_best=w;
       min_mse=MSE_Test;
   end
end
%% Test
=======================================================
===========
    E=0;
for ii=1:300
   x=Test_data(ii,:)';
   o2=h(w_best,x);
   e=Target_test(ii)-o2;
   E=E+e^2;
   yest(ii)=o2;
end
  E/300
  %%

  figure
  plot(yest,'--b','linewidth',2)
  hold on
  plot(Target_test,'--r','linewidth',2)
```

Snippet for calculating neural network output

```
function
[o1,o2,net1,net2]=FeedForward_NN(u,b1,b2,w1,w2)
%% Activation functions
f1 =@(x)( 8-exp(-x))./(1+exp(-x));
f2 =@(x)( 8-exp(-x))./(1+exp(-x));
%f2 =@(x) 1./(1+exp(-x));
%% Hidden Layer
u1=[b1;u];
net1=w1'*u1;
o1=f1(net1);
%% Out Layer
u2=[b2;o1];
net2=w2'*u2;
o2=f2(net2);
```

UKF snippet

```
function [x,P]=ukf(fstate,x,P,hmeas,z,Q,R,u_in)
% UKF   Unscented Kalman Filter for nonlinear dynamic
systems
% [x, P] = ukf(f,x,P,h,z,Q,R) returns state estimate,
x and state covariance, P
% for nonlinear dynamic system (for simplicity, noises
are assumed as additive):
%            x_k+1 = f(x_k) + w_k
%            z_k   = h(x_k) + v_k
% where w ~ N(0,Q) meaning w is gaussian noise with
covariance Q
%        v ~ N(0,R) meaning v is gaussian noise with
covariance R
% Inputs:    f: function handle for f(x)
%            x: "a priori" state estimate
%            P: "a priori" estimated state covariance
%            h: fanction handle for h(x)
%            z: current measurement
%            Q: process noise covariance
%            R: measurement noise covariance
```

```
% Output:    x: "a posteriori" state estimate
%            P: "a posteriori" state covariance
%
% Example:
%{
n=3;        %number of state
q=0.1;      %std of process
r=0.1;      %std of measurement
Q=q^2*eye(n); % covariance of process
R=r^2;        % covariance of measurement
f=@(x)[x(2);x(3);0.05*x(1)*(x(2)+x(3))];  % nonlinear
state equations
h=@(x)x(1);                               %
measurement equation
s=[0;0;1];                                % initial
state
x=s+q*randn(3,1); %initial state          % initial
state with noise
P = eye(n);                               % initial
state covraiance
N=20;                                     % total
dynamic steps
xV = zeros(n,N);          %estmate        % allocate
memory
sV = zeros(n,N);          %actual
zV = zeros(1,N);
for k=1:N
  z = h(s) + r*randn;                     %
measurments
  sV(:,k)= s;                             % save
actual state
  zV(k)  = z;                             % save
measurment
  [x, P] = ukf(f,x,P,h,z,Q,R);            % ekf
  xV(:,k) = x;                            % save
estimate
  s = f(s) + q*randn(3,1);                % update
process
end
for k=1:3                                 % plot
results
  subplot(3,1,k)
  plot(1:N, sV(k,:), '-', 1:N, xV(k,:), '--')
end
%}
% Reference: Julier, SJ. and Uhlmann, J.K., Unscented
Filtering and
% Nonlinear Estimation, Proceedings of the IEEE, Vol.
92, No. 3,
```

```
% pp.40 8-422, 2004.
%
% By Yi Cao at Cranfield University, 04/01/2008
%
L=numel(x);                                         %numer of
states
m=numel(z);                                         %numer of
measurements
alpha=1e-3;                                         %default,
tunable
ki=0;                                               %default,
tunable
beta=2;                                             %default,
tunable
lambda=alpha^2*(L+ki)-L;                            %scaling
factor
c=L+lambda;                                         %scaling
factor
Wm=[lambda/c 0.5/c+zeros(1,2*L)];                   %weights
for means
Wc=Wm;
Wc(1)=Wc(1)+( 8-alpha^2+beta);                       %weights
for covariance
c=sqrt(c);
X=sigmas(x,P,c);                                    %sigma
points around x
[x1,X1,P1,X2]=ut(fstate,X,Wm,Wc,L,Q,u_in);
%unscented transformation of process
% X1=sigmas(x1,P1,c);                                 %sigma
points around x1
% X2=X 8-x1(:,ones(1,size(X1,2)));
%deviation of X1
[z1,Z1,P2,Z2]=ut(hmeas,X1,Wm,Wc,m,R,u_in);
%unscented transformation of measurments
P12=X2*diag(Wc)*Z2';
%transformed cross-covariance
K=P12*inv(P2);
x=x1+K*(z-z1);                                      %state
update
P=P 8-K*P12';
%covariance update

function [y,Y,P,Y1]=ut(f,X,Wm,Wc,n,R,u_in)
%Unscented Transformation
%Input:
%         f: nonlinear map
%         X: sigma points
%        Wm: weights for mean
%        Wc: weights for covraiance
```

```
%        n: numer of outputs of f
%        R: additive covariance
%Output:
%        y: transformed mean
%        Y: transformed smapling points
%        P: transformed covariance
%       Y1: transformed deviations
f1 =@(x)( 8-exp(-x))./(1+exp(-x));
f2 =@(x)( 8-exp(-x))./(1+exp(-x));
L=size(X,2);
y=zeros(n,1);
Y=zeros(n,L);
for k=1:L
    Y(:,k)=f(X(:,k),u_in);
    y=y+Wm(k)*Y(:,k);
end
Y1=Y-y(:,ones(1,L));
P=Y1*diag(Wc)*Y1'+R;

function X=sigmas(x,P,c)
%Sigma points around reference point
%Inputs:
%       x: reference point
%       P: covariance
%       c: coefficient
%Output:
%       X: Sigma points

A = c*chol(P)';
Y = x(:,ones(1,numel(x)));
X = [x Y+A Y-A];
```

References

1. S.J. Julier, J.K. Uhlmann, New extension of the Kalman filter to nonlinear systems. AeroSense **97**, 182–193 (1997)
2. F. Gustafsson, G. Hendeby, Some relations between extended and unscented Kalman filters. IEEE Trans. Signal Process. **60**, 545–555 (2012)
3. E.A. Wan, R. Van Der Merwe, The unscented Kalman filter for nonlinear estimation, in *Adaptive Systems for Signal Processing, Communications, and Control Symposium 2000. AS-SPCC. The IEEE 2000* (2000), pp. 153–158
4. S.S. Haykin, *Kalman filtering and neural network* (Wiley Online Library, 2001)

Designing Neural-Fuzzy PID Controller Through Multiobjective Optimization

8

8.1 Introduction

PID controllers are commonly used to control industrial systems thanks to their simplicity and efficiency. These controllers are proportional–integral–derivative, each of which has its gains. If these gains are properly selected, they can obtain the proper solution from the system under control; however, if they are not properly selected, they lead to poor system performance and, as a result, to instability. The integrator term reduces the tracking error to zero, while the proportional and derivative terms improve stability. The derivative term is sensitive to changes in system output, which improves the stability of the closed loop. However, as previously stated, the important point is to properly select the controller's parameters. The parameters of this controller can be designed using either of the two methods listed below:

A. Classic methods, which are primarily based on experience.
B. Intelligent search engines such as GA and PSO.

8.2 Classic Methods

8.2.1 Ziegler–Nichols Method

This is one of the most widely used and practical methods in the industry. The process step response and its frequency response are used to determine the PID controller parameters in this method. The controller parameters can be regulated using the specifications derived

A. Mohammadzadeh et al., *Neural Networks and Learning Algorithms in MATLAB*,
Synthesis Lectures on Intelligent Technologies,
https://doi.org/10.1007/978-3-031-14571-1_8

from the process using these methods and the relevant table. It merits mention that the obtained table is based on experience.

8.2.2 Cohen-Coon Method

The following conversion function is used to model the process in this method:

$$G_p = \frac{k_p}{1 + sT} e^{-sL} \tag{8.1}$$

It is possible to obtain the controller parameters by modeling the process with kp, L, and T parameters, which represent the gain, latency, and time constant of the process, respectively.

As previously stated, the controller parameters are regulated using an experimentally obtained table in the methods described above. Obtaining the parameters is obviously not highly accurate and may not be suitable for complex systems. In general, the PID controller's parameters should be set so that the overshoot is low, the settling time is short, and the rise time is fast.

8.2.3 Smart Methods

Optimization in complex systems has been partially solved with the introduction of smart search methods such as GA and PSO. Evidently, classic optimization methods are primarily based on derivation. As the system dimensions expand, the optimization will improve. This method also has a high likelihood of being trapped in local minima as the system dimensions expand. Unlike classic methods, which move in a point-by-point manner, smart methods start the search with a set of points and then compare the objective functions of these points to find the proper solution after some iterations. When compared to classic optimization methods, the likelihood of being trapped in local minima is greatly reduced with these methods. When using smart methods to design PID controller parameters, these parameters are searched in the specified space in such a way that the objective function, which is essentially a measure of system error under control, is minimized. The objective function can differ depending on the desired objective, steady-state error or proportional offset error. The following objective functions, for example, could be taken into account:

$$Fitness = \int t \times e^2(t)dt \tag{8.2}$$

$$Fitness = \int t \times |e(t)|dt \tag{8.3}$$

where e (t) and t represent controlled system's error and time, respectively.

Or, if the criterion is the control signal applied to the intended process of the designer, the objective function can be selected as follows:

$$Fitness = \int (p \times e^2(t) + +q \times u^2(t))dt \tag{8.4}$$

where p and q are the designer's specified weights. And *u (t)* is the control signal.

As previously stated, the output of the controlled system should have low overshoot, low settling time, and fast rise time, which is the ideal goal when designing a PID controller. If one of the subjects is given unintentionally greater precision in the designing process, the others perform poorly. A more precise analysis of these characteristics can be presented in second-order linear systems. For example, in reducing the settling time, the overshoot increases unintentionally. It should be noted that because many higher-order linear systems can be approximated by a second-order system, such analyses of overshoot and settling time are correct.

The question that arises is whether solutions with low overshoot and settling time can be achieved simultaneously by minimizing the errors defined in the above equations as the objective function for optimization algorithms such as GA and PSO.

The answer to this question is most likely no. The reason for this is that, as previously stated, these two indicators are in conflict in terms of behavior, and by improving one, the other worsens. In such single-objective problems, reducing one of the indicators, such as the settling time, inevitably increases the system's overshoot, and vice versa. Obtaining PID controller parameters can thus be a multiobjective problem with optimization vectors that include overshoot, settling time, and rise time values. Furthermore, the technique required to solve a multiobjective problem differs significantly from standard optimization for a single-objective problem. As previously stated, multiobjective optimization is the search for optimizing the parameters of a target vector.

8.2.4 Single-Objective Optimization

The following is the definition of a single-objective optimization:

Identifying the vector $\vec{x}$ that optimizes the function $f(\vec{x})$ so that:

$$g_i(\vec{x}) \leq 0, \quad \mathrm{i} = 1, \ldots, \mathrm{m} \tag{8.5}$$

$$h_i(\vec{x}) = 0, \quad i = 1, \ldots, p \tag{8.6}$$

where $\vec{x}$ is the solution vector and is represented as $\vec{x} = [x_1, x_2, \ldots, x_r]^T$, and m is the number of inequality constraints and p is the number of equality constraints (constraints can be linear or nonlinear in both cases).

8.2.5 Multiobjective Optimization

The following is a definition of multiobjective optimization:

> A decision-making vector with multiple criteria that meets the constraints and optimizes a function vector whose elements represent the objective function. These functions provide a mathematical interpretation of efficiency criteria that are diametrically opposed. Thus, improvement denotes the discovery of a solution that can be considered acceptable for all objective functions.

The following is a representation of the above definition:

Finding the vector $\vec{x}^* = [x_1^*, x_2^*, \ldots, x_r^*]^T$ in such a way that the following m inequality constraints are observed:

$$g_i(\vec{x}) \leq 0, \quad i = 1, \ldots, m \tag{8.7}$$

And also p equality constraints are met as follows:

$$h_i(\vec{x}) = 0, \quad i = 1, \ldots, p \tag{8.8}$$

And also the following function vector is optimized:

$$\vec{f_i}(\vec{x}) = [f_1(\vec{x}), \ldots., f_k(\vec{x})]^T \tag{8.9}$$

where $\vec{x} = [x_1, x_2, \ldots, x_r]^T$ is the vector of decision variables.

In other words, it is intended to estimate a specific set as $x_1^*, x_2^*, \ldots, x_r^*$ among F sets (feasible region) that satisfy all of the conditions of (7.8) and (8.8) in order to yield optimal values for all k objective functions.

When there are multiple goals that must be optimized at the same time, multiobjective optimization methods are used. Consider the functions F_1 and F_2 shown in Fig. (8.2). The minimum values of these functions, as seen, are $F_1(0)$ and $F_2(1)$. There will be no single minimum value if it is intended to minimize both of these functions at the same time. For example, $x = 0$ is not a minimum value for the $F_2(x)$ function, so the value of $x = 1$ is also not a minimum value for the $F_1(x)$ function. In this case, the value of x can be a number in the [0, 1] range, which is referred to as an optimal solution. Thus, without default information about the problem and its application, none of the solutions take precedence over the others. In general, for the sake of simplicity, each of the objective functions is assumed to be about to be minimized. If there is an objective function, such as F_i, that should be maximized, it is actually equal to minimizing the $-F_i$ function (Fig. 8.1).

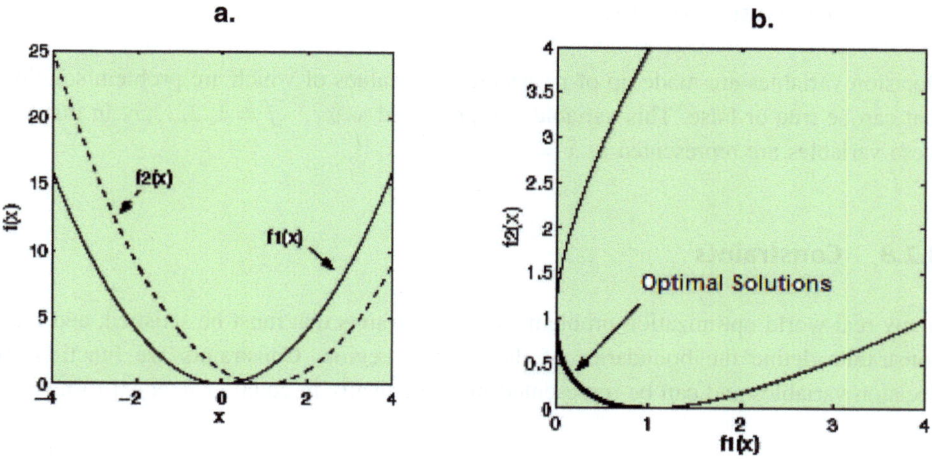

Fig. 8.1 **a** Optimization problem of dual goals. **b** Target space

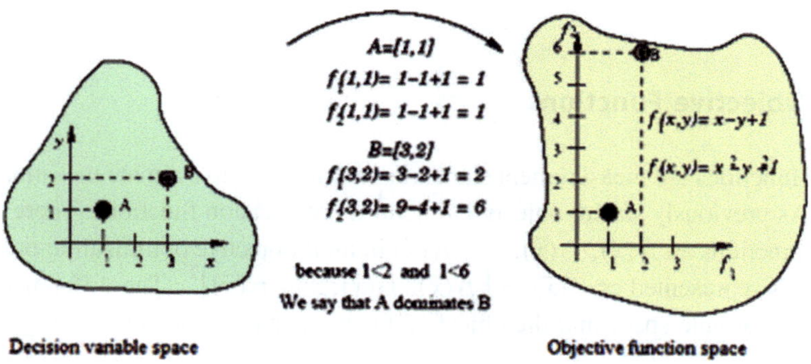

Fig. 8.2 Difference between target space (right) and decision space (left)

8.2.6 Primary Definitions

This section presents the most important definitions of optimization problems, as well as specific definitions of multiobjective optimization, and examines their roles in multiobjective optimization.

8.2.7 Decision Variables

Decision variables are made up of r variables, the values of which are problem solutions that can be true or false. This variable is represented as $x_j, \quad j = 1, 2, \ldots r$. In this text, these variables are represented as $\vec{x} = [x_1, x_2, \ldots, x_r]^T$.

8.2.8 Constraints

Many real-world optimization problems have constraints that must be satisfied, and these constraints define the boundaries of the feasible region. Constraints are functions of decision variables and can be represented mathematically in equal or unequal forms.

$$g_i(\vec{x}) \leq 0, \quad i = 1, \ldots, m \tag{8.10}$$

$$h_i(\vec{x}) = 0, \quad i = 1, \ldots, p \tag{8.11}$$

8.2.9 Objective Functions

Objective functions are measurement standards that are used to estimate the suitability of a solution. As previously stated, objective functions are decision functions. There are $k \geq 2$ objective functions as $f_1(\vec{x}), f_2(\vec{x}), \ldots, f_k(\vec{x})$ in multiobjective optimization problems. In this text, f is represented as $\vec{f}(\vec{x}) = \left[f_1(\vec{x}), f_2(\vec{x}), \ldots, f_k(\vec{x})\right]^T$. The difference between the decision variable space and the objective function space is depicted in Fig. (8.2).

8.2.10 Dominance

The decision vector $\vec{x}_1 \in F$ strictly dominates the decision vector $\vec{x}_2 \in F$ and is denoted as $\vec{x}_1 \prec \vec{x}_2$ if:

1. In all objectives, the decision vector $\vec{x}_1$ outperforms the $\vec{x}_2$ vector, or in other words:

$$f_i(\vec{x}_1) < f_i(\vec{x}_2), \quad \forall i = 1, \ldots, m \tag{8.12}$$

The decision vector $\vec{x}_1 \in F$ weakly dominates the decision vector $\vec{x}_2 \in F$ and is denoted as $\vec{x}_1 \preceq \vec{x}_2$ if:

1. In all objectives, the decision vector $\vec{x}_1$ performs poorly than the $\vec{x}_2$ vector, or in other words:

$$f_i(\vec{x}_1) \leq f_i(\vec{x}_2), \quad \forall i = 1, \ldots, m \tag{8.13}$$

2. The decision vector $\vec{x}_1$ was expressly better than the $\vec{x}_2$ vector in at least one of the objectives, or in other words:

$$f_i(\vec{x}_1) \leq f_i(\vec{x}_2), \text{At least for one}\, i = 1, \ldots, m \tag{8.14}$$

The decision vector $\vec{x}_1 \in F$ is indifferent toward the decision vector $\vec{x}_2 \in F$ and it is represented as $\vec{x}_1 \sim \vec{x}_2$ if:

$$\vec{x}_1 \leq \vec{x}_2 \wedge \vec{x}_2 \leq \vec{x}_1 \tag{8.15}$$

The following is a numerical representation of the aforementioned states:

$$\begin{aligned}
&\vec{\mathrm{x}}_1 = [2.4, 5.3, 4.5]^{\mathrm{T}}, \vec{\mathrm{x}}_2 = [2.4, 5.3, 4.8]^{\mathrm{T}}, \\
&\vec{\mathrm{x}}_3 = [3.4, 5.4, 4.7]^{\mathrm{T}} \\
&\vec{x}_1 \prec \vec{x}_3 \text{ because } \; 2.4 < 3.4, \quad 5.3 < 5.4 \\
&\text{and } 4.5 < 4.7 \\
&\vec{x}_1 \preceq \vec{x}_2 \;\; \text{because} \;\; 2.4 = 2.4, \quad 5.3 = 5.3 \\
&\text{and} \;\; 4.5 < 4.8 \\
&\vec{x}_2 \sim \vec{x}_3 \;\; \text{because} \;\; 2.4 < 3.4, \quad 5.3 < 5.4 \\
&\text{and} \;\; 4.8 > 4.7
\end{aligned} \tag{8.16}$$

8.2.11 Non-Dominated Set

The set of non-dominated solutions denoted by P' includes those solutions in the solution space that are not dominated by any of the elements of the set P Fig. (3.8) shows an illustration of this concept (Fig. 8.3).

8.2.12 Pareto Principle

It is not possible to find a single solution that is optimal for all objectives at the same time in multiobjective optimization. In this case, there are some optimal solutions such that improving one of their objective values would have a negative impact on at least

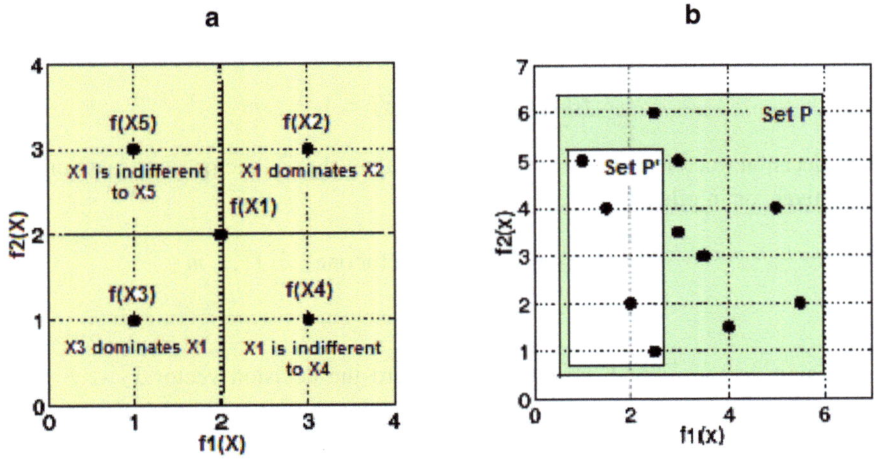

Fig. 8.3 **a** Objective space and examples of domination. **b** Non-dominated set P'

one other objective value. To put it another way, there is no way to improve an optimal solution's objective value without affecting other objective values in the solution vector. Edgeworth presented this definition in 1881, and the French-Italian sociologist and economist Vilfredo Pareto is credited with the final definition of this concept.

8.2.13 Optimal Pareto Solution

Decision vector $\vec{x}_1 \in F$ is called an optimal Pareto solution if there is no $\vec{x}_2 \in F$ to dominate it. If the corresponding decision vector of an objective function is an optimal Pareto solution, it is called the optimal Pareto solution.

Note: When all of the Pareto optimizations for a multiobjective optimization problem are found, the problem is mathematically solved.

8.2.14 Optimal Pareto Set

The optimal Pareto set, denoted P*, is defined as follows:

$$P^* = \left\{ \vec{x}_1 \in F | \neg \exists \vec{x}_2 \in F \,\middle|\, \vec{f}(\vec{x}_2) \preceq \vec{f}(\vec{x}_1) \right\} \tag{8.17}$$

In other words, if no other decision vector, such as $\vec{x}_2 \in F$ from the allowed F set, dominates it, the decision vector $\vec{x}_1 \in F$ from the allowed F set is an optimal Pareto solution. It is worth noting that the optimal Pareto set is defined in the space of decision variables.

When solutions are plotted in objective space, the set of non-dominated vectors is known as the Pareto front. The most common method for generating a Pareto front is to first generate a set of feasible F points and then the corresponding $f(F)$. When there are sufficient points, the Pareto front can be generated by locating and estimating non-dominated points.

8.3 Objectives of Multiobjective Optimization

- The primary goal of multiobjective optimization algorithms is to find a set of solutions in multiobjective problems that make a balance between different objectives. As previously stated, the goal is to find a set of non-dominated solutions known as the optimal Pareto set. This is accomplished by multiobjective optimization algorithms estimating the Pareto front. This includes the three objectives listed below:
- Minimizing the distance between the discovered solutions and the Pareto front.
- Maximizing the variety of discovered non-dominated solutions to represent more points belonging to the Pareto front.
- Maintaining discovered non-dominated solutions.

  The first goal can be met by developing an appropriate fitness function to assess the quality of the discovered solution for the various objectives. In general, fitness functions are classified into three types, which are as follows:
- **Aggregation-based method**, in which the fitness function is a weighted sum of the objective functions.
- **Criteria-based method**, in which the fitness calculation is switched between different objectives at different stages of optimization.
- **Dominance-based method**, in which the fitness is proportional to the dominance ranking of solutions. Many dominance-based sorting designs employ an archive of discovered non-dominated solutions as well.

Methods for maintaining the diversity of solutions have been proposed in order to achieve the second objective. These methods increase the likelihood of absorbing decision vectors into lower density areas of the Pareto front.

The third objective is achieved through the use of an archive of discovered non-defeated solutions, which is essentially an elitist strategy in which the most optimal discovered solutions are stored in the repository.

8.3.1 Common Algorithms in Solving Multiobjective Optimization

Different algorithms are used to solve multiobjective optimization problems. Among the proposed methods based on demographic algorithms are the following:

GA and PSO

The following are some of the genetic algorithm-based methods that have been introduced:

Srinivas and Deb proposed the non-dominated sorting genetic algorithm (NSGA-II algorithm) in 1994.

Horn and Nafpliotos developed the niched Pareto genetic algorithm (NPGA algorithm)

Strength pareto evolutionary algorithm (SPEA algorithm) was proposed by Zitzler and Thiele.

Among the introduced PSO-based algorithms, the following are worth mentioning:

The adaptive weighted particle swarm optimization (AWPSO) algorithm, which is an improved version of the PSO algorithm, aims to improve the efficiency of this exploratory optimization method.

8.4 Designing Multiobjective PID Controller

This section describes a new approach to designing PID controllers that is based on a multiobjective BMPSO algorithm. As previously stated, errors are variously used as fitness functions of optimization algorithms in common smart methods of designing PID controllers. Furthermore, simultaneous multiobjective optimization for PID controllers was not possible in these algorithms, and valuing one made the other worse. Thus, in controllers designed using single-objective methods, only one solution is obtained, based on which only one of the aforementioned components, namely overshoot or settling time, can be controlled. The obtained solution satisfies only one of these components, not both.

Based on the foregoing, it is essential to oppose a new method for designing PID controllers in order to achieve better results. As a result, multiobjective optimization algorithms appear to be appropriate. The reason for this is that multiobjective optimization algorithms, in addition to providing a diverse set of solutions for the system designer to use based on various conditions, also allow for simultaneous optimization of the aforementioned points. Therefore, unlike single-objective optimization, in this method, there are sets of solutions that have no precedence over the others and are dispersed in solution space according to the value of each of the objectives.

The introduction of objective functions is the most important action taken to design a PID controller using the multiobjective optimization method. As previously stated, this algorithm's objective functions include settling time, rise time, and overshoot. The important point here is that the objective function be selected in such a way that it is compatible with the general concept of multiobjective optimization in the contradiction of objectives. Overshoot and settling time, for example, are contradictory in most control systems and can be considered as two objectives in designing PID controllers in this method. The desired objective can thus be achieved by defining the following two objective functions, the purpose of which is to maximize them.

1. In order to optimize the settling time

$$F_1 = \frac{1}{1 + S_t} \tag{8.18}$$

2. In order to optimize the overshoot

$$F_2 = \frac{1}{1 + O} \tag{8.19}$$

where O and St are the overshoot values and the system output's normalized settling time, respectively. PID controller is designed as multiobjective by simultaneous maximization of two objective functions in (8.18) and (8.19) in multiobjective optimization, and a set of solutions is yielded.

The space of the problem decision variables in this method is the same as the space of the PID controller design parameters, i.e. KP, KI, and KD, and the problem's objective space is two-dimensional and includes the defined functions F1 and F2. After introducing the above for the BMPSO algorithm, the proposed method begins optimizing the objective functions simultaneously, and in the end, it presents a set of solutions that are distributed along the Pareto front of the problem. What is important to note here is the variety of discovered solutions based on this method, which allows the controller designer to select one of the solutions based on the need and significance of the objective functions. This method of designing a PID controller is known as multiobjective PID (MOPID). The practical results of using this method in designing PID controllers for frequency load control of power systems are discussed next.

8.5 Designing a MOPID Controller for a Sample Power System

The method of multiobjective optimization was described previously. It was pointed out that this method can be used to design the parameters of a PID controller in order to achieve the desired results. For this optimization method, two objective functions that should be maximized were defined. These objective functions included overshoot and settling time values. Indeed, as these objective functions are maximized, the desired goals of control systems, namely overshoot and settling time, are attained.

As mentioned in previous chapters, the goal of frequency load control is to restore the regions' frequencies and transmission power to their nominal values. Thanks to their simplicity and high capability, PID controllers are commonly used in industrial system controls, such as frequency load. In this regard, what is challenging in large systems is obtaining the parameters of this controller, and their proper regulation can be crucial.

The design methods of this controller were described in the previous sections, and they were divided into two categories: classic and smart. Classic methods rely on personal experiences and trial and error. The parameters of the PID controller are obtained in smart methods by defining an objective function for evolutionary algorithms, which is a standard for the system under control. It was also stated that obtaining a solution with appropriate overshoot and settling time is very favorable, so that these two components are in conflict and can be presented as a multiobjective problem.

The multiobjective optimization method is used in this section to design PID controller parameters for the frequency load problem. Figure 8.12 of the first chapter depicts the power system used for simulation. Appendix represents the system's dynamic equations as well as the values of its nominal parameters. The system output (ACEi) is used to calculate the overshoot and settling time for each control area, as shown in Eq. 8.12. After defining the overshoot and settling time values for the multiobjective optimization problem, the next step is to normalize them. Only the settling time is normalized in simulation because the overshoot value is roughly between 0 and 1. If the simulated time is 10 s, the actual settling time is divided by that and the normalized value is obtained.

Some states of Power Exchange (PX) and changes in system operating points are used for simulation to describe the performance and capability of the optimized algorithm.

8.5.1 First State

In this scenario, the power system is operating at its nominal point, and the GPM matrix defined in Eq. 8.18, which represents power exchange among regions, is taken into account as follows.

$$GPM = \begin{bmatrix} .5 & .5 & 0 & 0 \\ .5 & .5 & 0 & 0 \\ 0 & 0 & .5 & .5 \\ 0 & 0 & .5 & .5 \end{bmatrix}$$

This matrix indicates that the loads demanded by the power system Discos must be supplied by the region Gencos. As discussed earlier, the columns of this matrix represent Discos and the rows represent Gencos of the power system. For instance, the demanded load of the first Disco is supplied equally by the first and second Gencos, which are in the region itself. If the Discos' demanded load (ΔP_{Li}) is assumed to be equal and equal to PU 1, the output of the Gencos for each region is as follows:

$$\Delta P_{Mi} = gpf_{i1} \times \Delta P_{L1} + gpf_{i2} \times \Delta P_{L2} + gpf_{i3} \times \Delta P_{L3} + gpf_{i4} \times \Delta P_{L4}$$

Thus,

$$\Delta P_{M1} = .5 \times \Delta P_{L1} + .5 \times \Delta P_{L2} + 0 \times \Delta P_{L3} + 0 \times \Delta P_{L4} \\ = .1\ PU$$

Similarly, the following can be written for other Gencos:

$$\Delta P_{M2} = .1\ PU$$
$$\Delta P_{M3} = .1\ PU$$
$$\Delta P_{M4} = .1\ PU$$

The nominal value of the frequency and transmitted power change as these loads enter the regions, and these changes must be reduced to zero by the designed MOPID controller.

The population size for the BMPSO algorithm is 30 particles, and the number of running algorithm steps is 50. After running the algorithm, the generated Pareto front is as Fig. 8.4.

This figure depicts the obtained PID solutions in Pareto front for the two intended objective functions, namely overshoot value (F2) and settling time (F1). For the first state, 8 PID controllers for the BMPSO algorithm were obtained that are comparable due to overshoot and settling time values. Each coordinate in F1 and F2 spaces, indicated by o, was generated as a result of the power system's use of a PID controller. In fact, this is the dominance point of a multiobjective optimization algorithm, which generates corresponding PID controller parameters based on overshoot and settling time values. Thus,

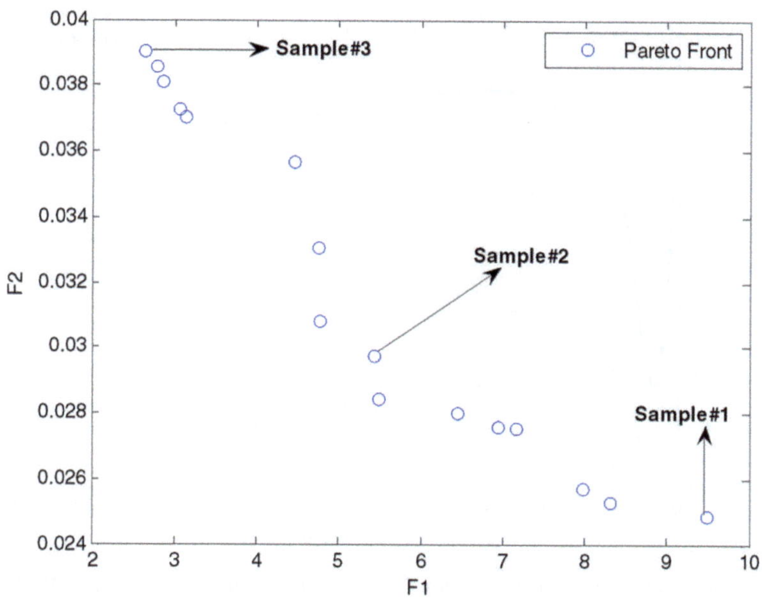

Fig. 8.4 The generated Pareto front for the first state

the controller's designer can use their corresponding controller parameters by prioritizing each of the objectives based on the obtained Pareto front.

Three of these coordinates are considered (*Sample* #1 ~ 3) for further explanation, and frequency changes in the two regions are discussed by entering their corresponding PID parameters in the power system. The BMPSO algorithm yielded the following parameter coefficients for the PID controller for the three samples:

$$\text{For } Sample\ \#1(K_P, K_I, K_D) = (3.9106, 0.56431, 3.9409):$$
$$\text{For } Sample\ \#2(K_P, K_I, K_D) = (3.8433, 0.09871, 3.1457):$$
$$\text{For } Sample\ \#3(K_P, K_I, K_D) = (3.6462, 0.049726, 2.3171):$$

The frequency changes in the first and second regions for these three examples are depicted in Figs. 8.5 and 8.6, respectively.

According to the system solution results and also while considering the obtained Pareto front, PID controllers obtained by the BMPSO algorithm have a wide range of values for the intended objectives. Given the obtained results, the desired PID control parameters in terms of evaluating each of the defined objectives can be extracted. When applied to a power system, for example, the PID controller corresponding to *Sample* #1 has a low overshoot and a high settling time when compared to others. Or, while the PID controller

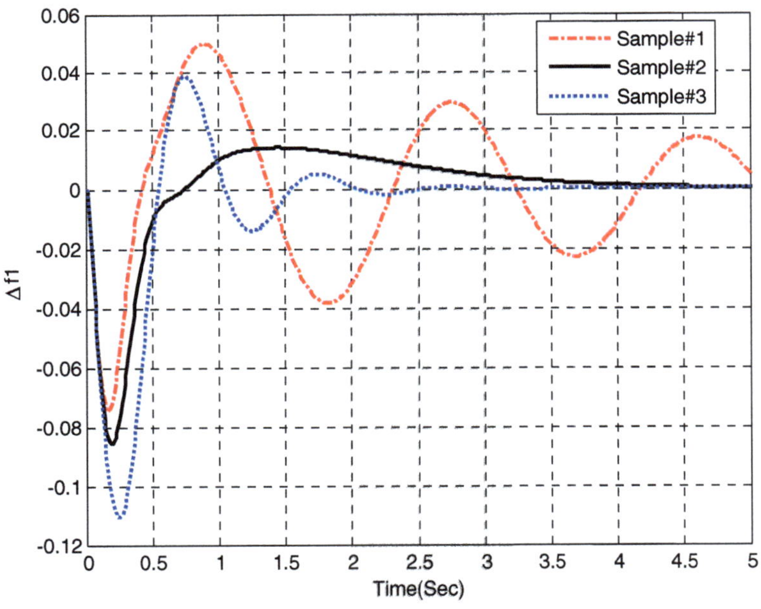

Fig. 8.5 Frequency changes in the first region for three PID samples obtained by BMPSO algorithm for the first mode

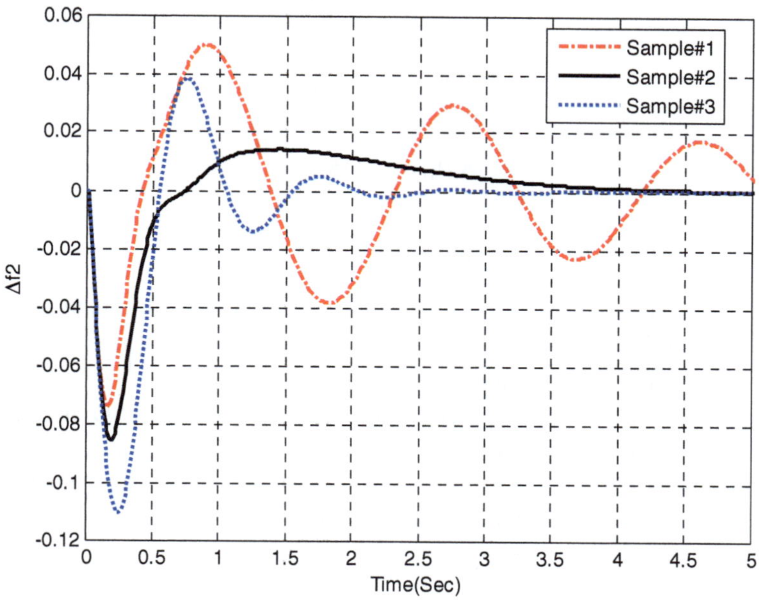

Fig. 8.6 Frequency changes in the second region for three PID samples obtained by BMPSO algorithm for the second mode

corresponding to *Sample* #3 has a low settling time in comparison to other solutions, it has a higher overshoot. One of these solutions (PID controllers) can be applied to the system based on the designer's preference in this controller and the characteristics of the system under control.

8.5.2 Second State

In this state, the nominal value of the power system's operating point changes, and the operating point's parameters, namely (Bi T12, Tpi), change to non-nominal values (25,0.435,0.35) (i = 1: 4). Furthermore, the GPM matrix is defined as follows, taking into account the power convention in the power system's Discos and Gencos:

$$GPM = \begin{bmatrix} .5 & .25 & 0 & .3 \\ .2 & .25 & .3 & 0 \\ 0 & .25 & .2 & .7 \\ .3 & .25 & .5 & 0 \end{bmatrix}$$

The generation of Gencos in each region should meet the following values, based on this matrix and the 0.1 PU load demand of Discos:

$$\Delta P_{M1} = .105\ PU$$

$$\Delta P_{M2} = .075\ PU$$

$$\Delta P_{M3} = .115\ PU$$

$$\Delta P_{M4} = .105\ PU$$

Given these values, the load flow from the first to second states should be −0.02 PU.

In this state, just as the first state, the population size for the BMPSO algorithm is 30 particles, and the number of running algorithm steps is 50.

The frequency load in power systems was obtained by the BMPSO algorithm, which was similar to the first state in that some of the PID controller parameters were obtained based on the values given to the defined objectives. For example, the parameters obtained for the PID controller for the three modes shown on the Pareto front ($Sample\ \#1 \sim 3$) are as follows:

$$\text{For } Sample\ \#1(\mathrm{K_P}, \mathrm{K_I}, \mathrm{K_D}) = (4.62, 0.62, 2.583):$$
$$\text{For } Sample\ \#2(\mathrm{K_P}, \mathrm{K_I}, \mathrm{K_D}) = (4.82, 0.255, 2.021):$$
$$\text{For } Sample\ \#3(\mathrm{K_P}, \mathrm{K_I}, \mathrm{K_D}) = (4.890, 1083, 1.27):$$

Furthermore, the frequency changes in the first and second states for these three PID controllers are depicted in Figs. 8.7 and 8.8, respectively.

The obtained PID controllers differ in terms of overshoot and settling time values, as shown in the figures. By simulating solutions, such results can be obtained for other controllers obtained in the Pareto front. Unlike the PID controllers described in the previous sections, this algorithm, the BMPSO algorithm, produces a set of solutions for the PID controller that differ in terms of valuing the objectives. This opens up a plethora of possibilities for designers in terms of extracting controller parameters based on the limitations of the system under control. It should be noted that no solution has priority over the others in this set of solutions, and they can be chosen based on the values assigned to each of the defined objectives.

8.6 Using Fuzzy-Neural Network for Gain Schedule

It was explained in previous chapters and sections that the three parameters of the power system (B_i, T_{12}, T_{pi}) are dependent on the system's operating conditions and may change depending on the system's operating conditions. As these parameters change, the designed controller should be able to deal with them. As a result, controllers with fixed structures

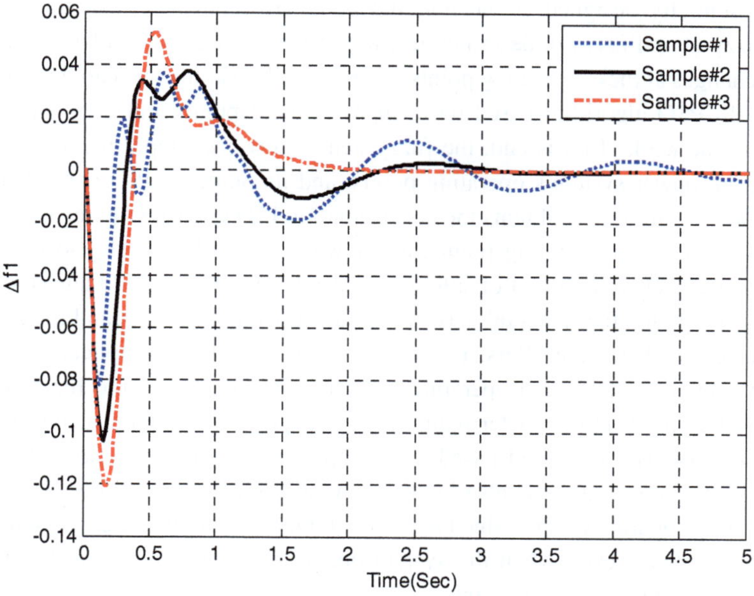

Fig. 8.7 Frequency changes in the first region for three PID samples obtained by BMPSO algorithm for the first mode

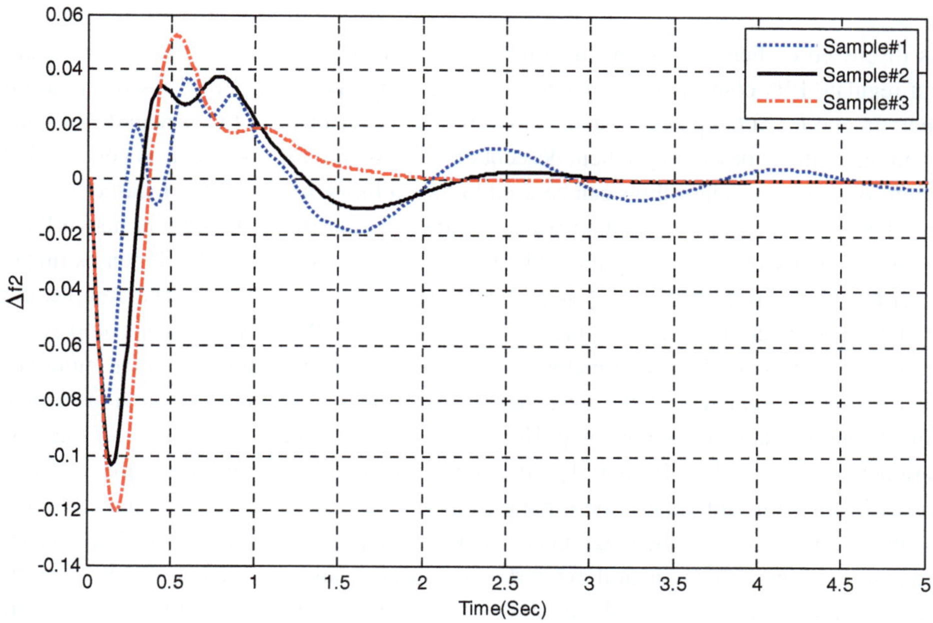

Fig. 8.8 Frequency changes in the second region for three PID samples obtained by BMPSO algorithm for the second mode

that are essentially designed at the nominal point are ineffective. For this reason, the Sugeno fuzzy-neural network described in Chap. 5 can be used to regulat PID controllers based on changes at these operating points. PID controller parameters can thus be obtained using the BMPSO algorithm based on changes at these points and then used to train a fuzzy-neural network. To this end, the fuzzy-neural network's input includes parameters indicating the power system's operating points, and its output is PID controller parameter. The BMPSO algorithm obtains the corresponding PID controller parameters for some of the power system's operating points, as shown in Table 8.1. It should be noted that designers can select from the set of solutions generated by the BMPSO algorithm for PID controllers. When the fuzzy-neural network is trained using this input and the corresponding parameters of PID controllers, it can be used in a real-time application to regulate controller parameters based on operating point monitoring. Using the fuzzy-neural network trained as the PID parameter regulator, an adaptive controller structure is obtained that can be quite well regulated based on the operating points of parameters. In classic control systems, this strategy is also known as a gain schedule controller. The difference is that the fuzzy-neural system trained based on Table 8.1 can effectively regulate the PID controller parameters even when the system operating points are not in this Table. The classic ones had some issues with this.

8.7 Fuzzy-Neural Network Training for PID Controller Regulation

As discussed earlier, it is intended to train a fuzzy-neural network in such a way that it can regulate PID controller parameters by monitoring operating points based on various operating points at the power system, and thus achieve an adaptive fuzzy-neural controller. Training methods based on gradient descent or smart search methods such as genetic algorithms or consensus algorithms can be used to train a fuzzy-neural network based on Table 8.1. The genetic algorithm method is used to train the desired fuzzy-neural network due to the simplicity of the training method based on smart algorithms. The Sugeno structure described in Chap. 5 is the fuzzy-neural network used in this section. The complexities of derivation operations are not present in this method. Moreover, when compared to gradient descent-based training methods, the likelihood of colliding with local minima is reduced. The parameters of the introduction and sequence sections, as previously mentioned, can be adapted and regulated. However, because the range of changes of the input parameters, namely T_{12}, B_i, and T_{pi}, is determined by better determining the values of mij and δ_{ij} as constant values, these parameters do not need to be trained, and a constant membership function for the fuzzy-neural network's input is designed. Instead, the parameters of the Sugeno fuzzy-neural network's output layer are regulated so that they can be trained based on Table 8.1. Without taking into account the regulating characteristics, the membership functions for these three inputs are defined as shown in Fig. 8.9.

Table 8.1 Power system operating points and corresponding PID parameters

Pattern	Operating conditions			PID parameters		
	T_{pi}	T_{pi}	T_{12}	K_P	K_I	K_D
1	10	0.145	0.125	3.16	0.3	4.59
2	10	0.145	0.275	4.3	0.256	3.91
3	10	0.145	0.425	2.3	0.150	2.64
4	10	0.345	0.125	3.16	0.08	1.58
5	10	0.345	0.275	4.52	0.534	2.55
6	10	0.345	0.425	3.84	0.06	1.07
7	10	0.545	0.125	2.9	0.45	1.80
8	10	0.545	0.275	3.4	0.25	2.02
9	10	0.545	0.425	3.8	0.20	1.92
10	10	0.145	0.125	4.98	0.16	1.75
11	20	0.145	0.125	4.91	0.11	0.75
12	20	0.145	0.275	4.18	0.07	1.18
13	20	0.145	0.425	3.6	0.12	1.03
14	20	0.345	0.125	2.1	0.18	0.33
15	20	0.345	0.275	3.53	0.102	1.29
16	20	0.345	0.425	4.3	0.16	1.87
17	20	0.545	0.125	3.6	0.04	0.08
18	20	0.545	0.275	4.02	2.07	4.59
19	20	0.545	0.425	4.08	0.156	3.91
20	20	0.145	0.125	1.12	0.20	4.44
21	30	0.145	0.125	3.67	0.404	3.707
22	30	0.145	0.275	3.08	0.070	1.697
23	30	0.145	0.425	3.04	0.16	0.37
24	30	0.345	0.125	2.534	0.57	2.23
25	30	0.345	0.275	3.286	11	3.04
26	30	0.345	0.425	2.1	0.068	4.9
27	30	0.545	0.125	4.25	1.124	1.081

In the sequential part of the Sugeno structure, the parameters that can be regulated are w_0^m, w_1^m, w_2^m, and w_3^m. As previously stated, each region of the fuzzy-neural network has three inputs and outputs. The initial population in the genetic algorithm is 20, the mutation and shear coefficients are 0.5 and 0.7, respectively, and the number of iterations is 150. Following training of the fuzzy-neural network, the structure shown in Fig. 8.8 can be used to regulate the PID controller based on changes in the operating point. This structure is an

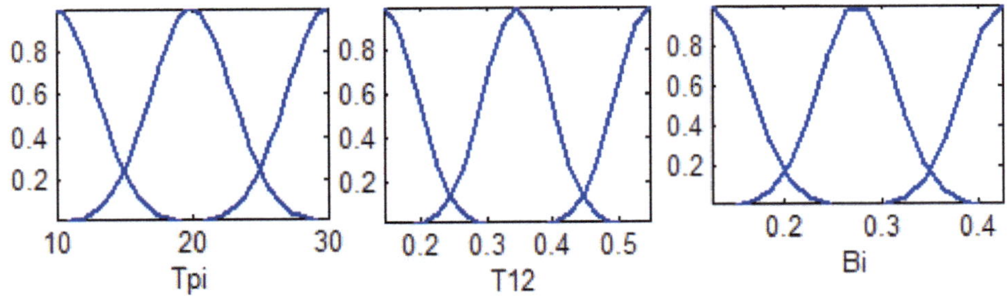

Fig. 8.9 Membership functions for fuzzy-neural network input

adaptive and nonlinear controller that can effectively solve load and frequency problems in power systems.

8.7.1 Simulation for Fuzzy-Neural Controller of Gain Schedule

First state **:**

The first state of load conversion and the system's nominal points, as discussed above, are taken into account using the structure shown in Fig. 8.8. The simulation results for this state are presented, which include changes in frequencies of the two regions in Figs. 8.10 and 8.11. The changes in transmission power that should be zero are also shown in Fig. 8.12.

Based on the simulation results, the fuzzy-neural network was able to regulate the parameters of the PID controller and achieve the objectives of the frequency load problem.

Second state **:**

In this case, the power transmission is considered as the second state but the variable power system parameters, namely B_i, T_{12}, Tpi, change from their non-nominal values to their nominal values (27,0.435,0.35), respectively. Frequency changes in two regions and transmission power are depicted in Figs. 8.13, 8.14, and 8.15, respectively.

The simulation results show that the objectives of frequency load are well achieved in a non-nominal point by proper regulation of the PID controller by the fuzzy-neural network.

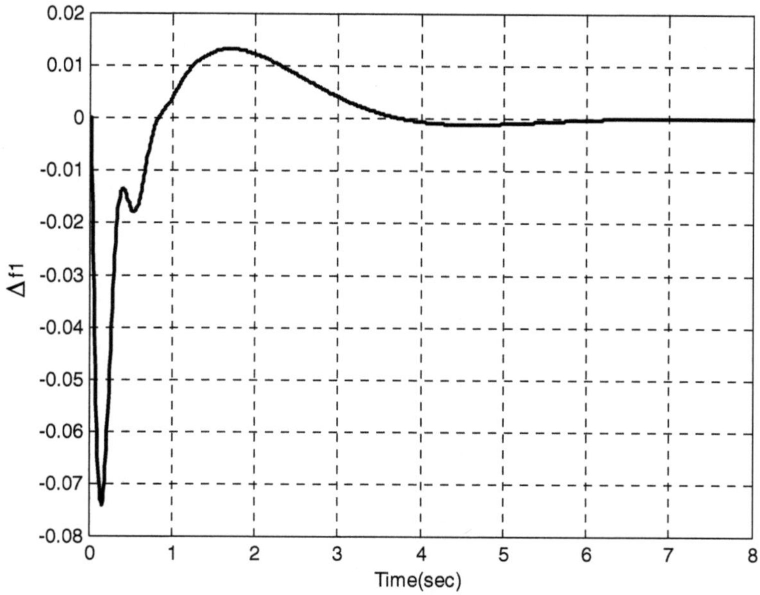

Fig. 8.10 Frequency changes in the first region for fuzzy-neural controller of the gain schedule

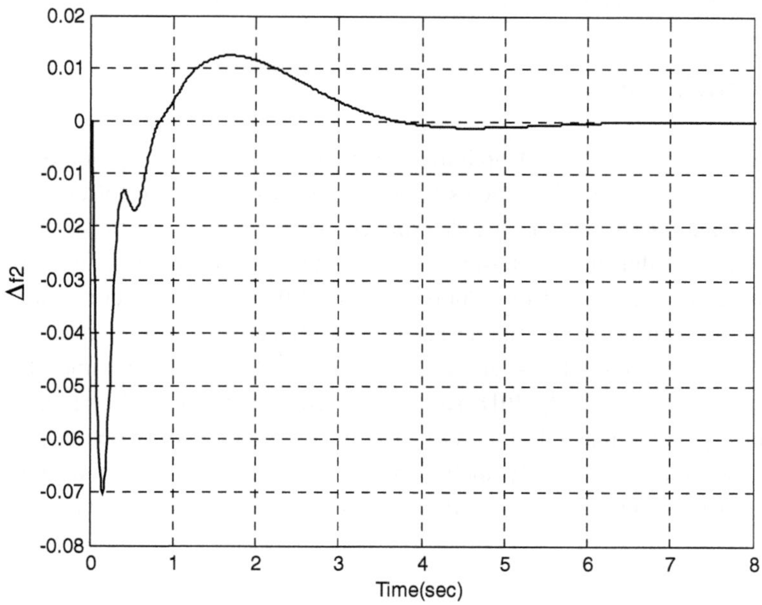

Fig. 8.11 Frequency changes in the second region for fuzzy-neural controller of gain schedule

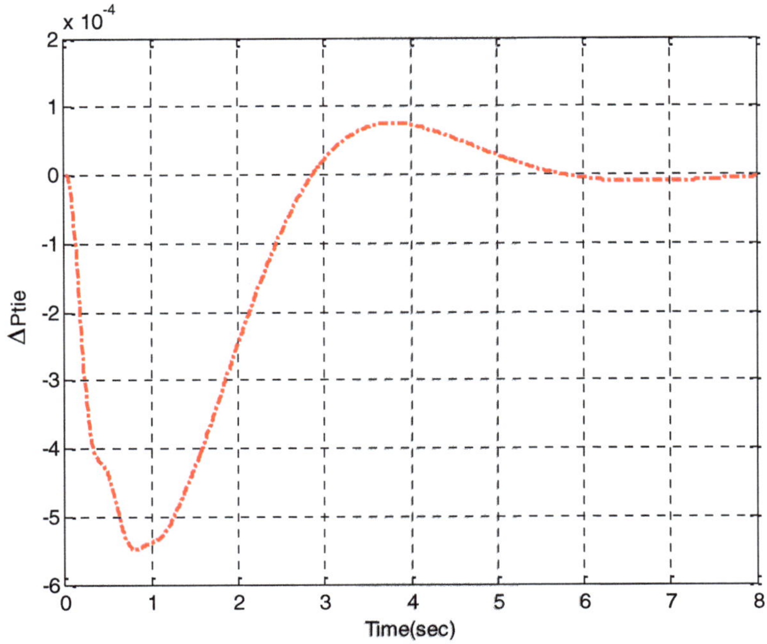

Fig. 8.12 Changes in transmission power for fuzzy-neural controller of the gain schedule

8.8 Conclusion

This chapter used the BMPSO optimization algorithm to design a multiobjective PID controller for frequency load problems in power systems. The values of overshot and settling time of the power system were determined as objective functions for multiobjective optimization algorithm in this design method. After running the algorithm for a power system operating point, a set of solutions for the PID controller were obtained, which can be discussed in terms of valuing each of the objectives. This solution set takes no precedence over other solutions, and each was obtained based on the importance of the determined objectives. Finally, PID parameters were obtained for all operating points of the system using this method and were trained to a Sugeno fuzzy-neural network as a gain schedule. In terms of timely application and observation of system operating points, this trained network can regulate the PID controller parameters quite well.

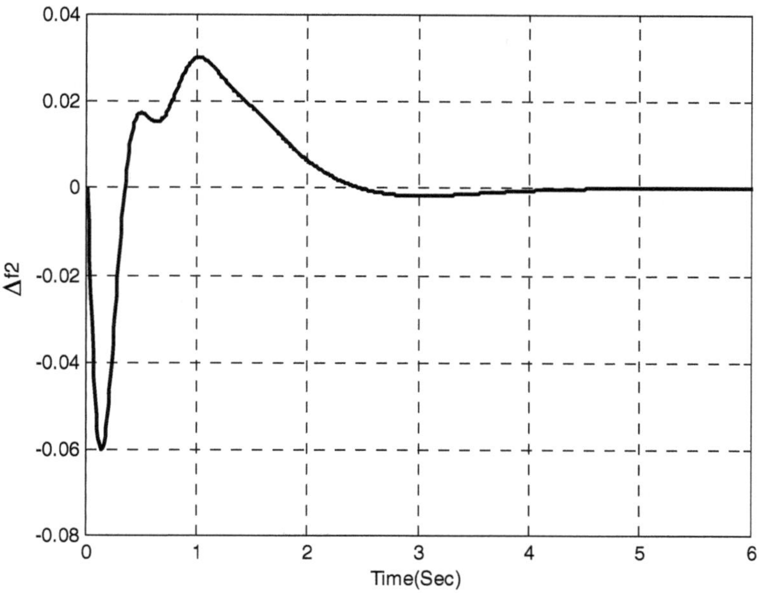

Fig. 8.13 Frequency changes in the second region for fuzzy-neural controller of gain schedule (second state)

8.9 Implementation in MATLAB

8.9.1 Dynamic Model of Power System

The operation of traditional load–frequency control has undergone transformation as a result of the power industry's new propensity toward structure renewal and competition. Frequency load control has evolved into a tradable product in the exterior space. However, the objectives of frequency load control problems have not changed, i.e., system frequency stability at nominal value, stabilizing current power of transmission lines in regions during schedule. In this structure, generators are referred to as generation companies (Genco), while consumers are referred to as distribution companies (Disco), which are tasked with purchasing power for loads. Additionally, in a competitive market, Gencos and Discos carry out power equations under the supervision of an independent system operator (ISO).

The following are some of ISO's responsibilities:

1. Calculating the amount of power needed by the Discos for LFC.
2. Keeping track of the amount of power purchased from Gencos.
3. Overseeing the cost of purchasing power from Gencos for the Discos.

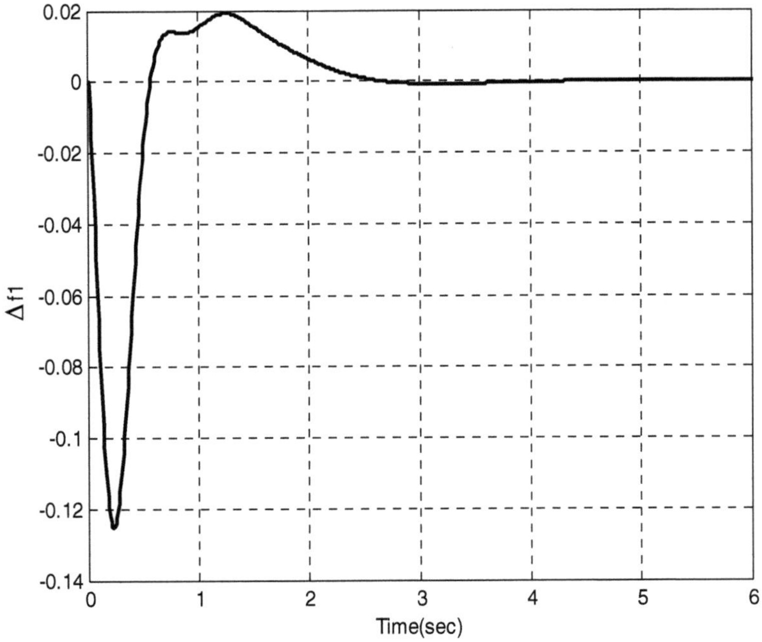

Fig. 8.14 Frequency changes in the first region for fuzzy-neural controller of the gain schedule (second state)

In systems with renewed structure, in addition to Gencos and Discos, transmission companies (Transcos) are responsible for transmitting the purchased power from Gencos to Discos.

Some Gencos may not participate in carrying out the frequency load control problem in this structure, whereas Discos can freely supply their needs with Gencos in their regions or other regions; there is a freedom policy in this regard. Therefore, Gencos and Discos may enter into a variety of contracts. For the sake of clarity, all trade is governed by ISO.

The generation participation matrix (GPM) is introduced to express this multi-faceted power exchange based on the power exchange between Gencos and Discos on the control regions. The participation factor of each Genco in the control region is represented by this matrix, and each control region has a number of Discos. This matrix's row corresponds to the number of Gencos and its column to the number of Discos. For example, the GPM matrix for a two-region power system depicted with four Discos and four Gencos is like this:

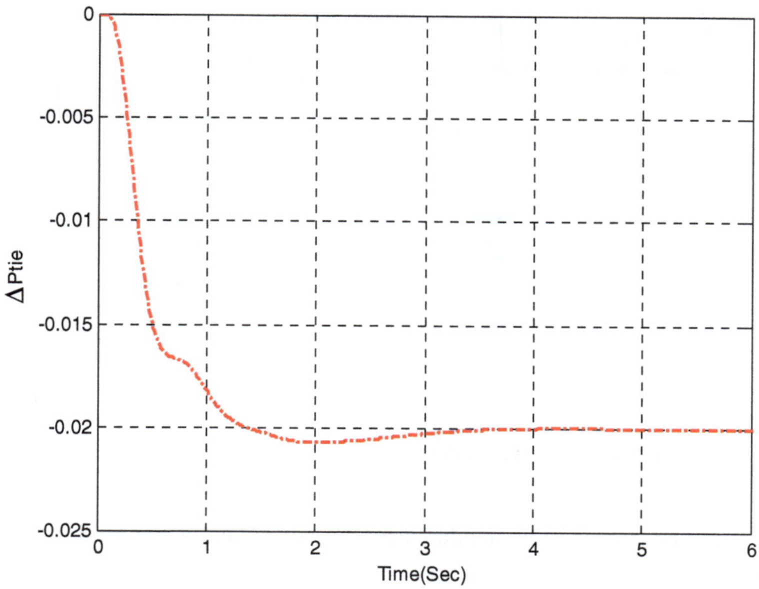

Fig. 8.15 Changes in transmission power for fuzzy-neural controller of the gain schedule (second state)

$$GPM = \begin{bmatrix} gpf_{11} & gpf_{12} & gpf_{13} & gpf_{14} \\ gpf_{21} & gpf_{22} & gpf_{23} & gpf_{24} \\ gpf_{31} & gpf_{32} & gpf_{33} & gpf_{34} \\ gpf_{41} & gpf_{42} & gpf_{43} & gpf_{44} \end{bmatrix} \tag{8.18}$$

As gpfij denotes the "generation participation factor," it is the coefficient of participation of the ith Genco in following (estimating) the demanded load of the jth Disco (according to their contract). The sum of all elements in each column in this matrix is one. Thus (Fig. 8.16):

$$\sum_{i=1}^{4} gpfij = 1 \tag{8.19}$$

Each element in the GPM matrix based on the load contract with Disco should be reflected on the control region and the reflected generation unit and met by them. This new information signal was not in the traditional frequency load control structure, which was not described in the preceding section ($P_{ref}\Delta$). This signal represents the power of the contract concluded between Discos and Gencos in the region or other regions, and the respective Genco must eventually meet the contract power. Besides, the allowed rate for current power from transmission lines should be calculated based on region contracts,

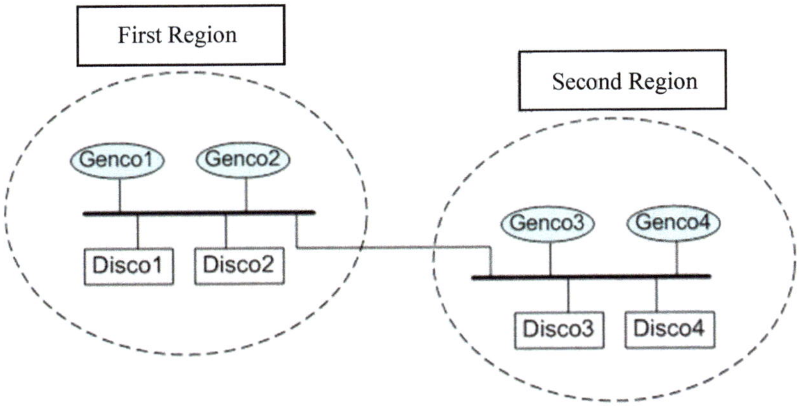

Fig. 8.16 Two-region renewed power system

with the difference between this regulated power and its real value used to calculate the ACE signal.

Based on the description of the restructured system, the dynamic modeling of this system for the frequency load control problem with some modifications is presented by adding new said signals in the traditional structure shown in Fig. 8.9.

A dotted line represents the new information signal in this figure. Moreover,

$_{ij}\alpha$: represents the ACE participation coefficient, which determines the participation of generators in supplying non-contractual load to the area. For each region the sum of $_{ij}\alpha$ is one.

ΔP_{Li}: The sum of load demand of Discos in each region.

ΔP_{TOTi}: Sum of the contractual loads inflowed to the *i*th region. (which equals the sum of Disco loads in the *i*th region).

ΔP_{di}: Sum of the noncontractual loads inflowed to the *i*th region.

$\Delta P_{tie\ 8\text{–}2}$, scheduled: the rate of transmitted power between two regions that should be conducted in the final state is calculated as follows, using the GPM matrix and the load demand of Discos:

$$\begin{aligned}\Delta P_{tie\ 8-2,\text{scheduled}} &= \sum \text{sum of exported power from the region} \\ &\quad - \text{sum of imported power to the region}) \\ &= \sum_{i=1}^{2}\sum_{j=3}^{4} gpf_{ji}\Delta P_j - \sum_{i=1}^{2}\sum_{j=3}^{4} gpf_{ji}\Delta P_i\end{aligned}$$

for the error of transmitted power between regions, it can be said:

$$\Delta P_{\text{tie 8 - 2, error}} = \Delta P_{\text{tie 8-2, actual}} - \Delta P_{\text{tie 8 - 2, scheduled}} \tag{8.20}$$

in light of this problem, the ACE signal for each region is calculated as follows, regardless of inter-region losses:

$$\begin{aligned} ACE_1 &= \Delta P_{\text{tie 8-2, error}} + B_1 \Delta f_1 \\ ACE_2 &= \beta \Delta P_{\text{tie 8-2, error}} + B_2 \Delta f_2, \quad \beta = -1 \end{aligned} \tag{8.21}$$

In fact, if the values of ACE1 and ACE2 are reduced to zero, the frequency load problem's objectives are met.

8.9.2 First Example

Assume that the demanded load of Discos (ΔP_L) in Fig. 8.12 is PU 0.1 and that.

The GPM matrix for power exchange contract is as follows:

$$GPM = \begin{bmatrix} .3 & .6 & .5 & .5 \\ .2 & .1 & .3 & .3 \\ .5 & .2 & .2 & .1 \\ 0 & .1 & 0 & .1 \end{bmatrix} \tag{8.22}$$

Given the load demanded by Discos and the GPM matrix, the following can be considered for transmitted power from the first to the second region, as well as load generated by Gencos.

$$\begin{aligned} &\Delta P_{\text{tie1-2, scheduled}} = \\ &[.5 \times (.1) + .3 \times (.1) + .5 \times (.1) + .3 \times (.1)] - [.5 \times (.1) \\ &+ 0 \times (.1) + .2 \times (.1) + .1 \times (.1)] \\ &= 0.08\,\text{pu} \end{aligned}$$

which denotes the rate at which power flows from the first to the second region.

In the first and second regions, the rate of power generated by generators based on the GPM matrix is as follows:

$$\Delta P_{Mi} = gpf_{i1} \times \Delta P_{L1} + gpf_{i2} \times \Delta P_{L2} + gpf_{i3} \times \Delta P_{L3} + gpf_{i4} \times \Delta P_{L4}$$

Production rate of generators (Gencos) in the first region:

$$\Delta P_{M1} = (.3 + .6 + .5 + .5) \times .1 = 0.19\,pu$$

$$\Delta P_{M2} = (.2 + .1 + .3 + .3) \times .1 = 0.09\,pu$$

Production rate of generators (Gencos) in the second region:

$$\Delta P_{M3} = (.5 + .2 + .2 + .1) \times .1 = 0.1\ pu$$

$$\Delta P_{M4} = (0 + .1 + 0 + .1) \times .1 = 0.02\ pu$$

The above values should be supplied by Gencos in each region so that the transmitted power from the first region to the second region eventually reaches the regulated value (pu.0.08) and frequency changes in regions reach zero.

8.9.3 Supplementary Ideas on Modeling the Power System for the Frequency Load Problem

Many complexities of the model were overlooked in previous chapters in order to model the power system frequency load problem, and a linear model for turbine and governor were used. The synchronizer coefficient was also calculated through linearization. Moreover, since the system's operating points have changed, B_i, which represents the frequency orientation coefficient, can be variable.

Thus, T_{ij}, B_i, and T_{pi} (synchronizer coefficient, frequency orientation coefficient, and power system time constant) were considered variables for modeling power systems based on changes in system operating conditions that could be changed within a certain range. Moreover, given the generation rate constraint (GRC), a nonlinear model was used for the problem statement in turbine modeling.

Based on the discussions presented in the modeling power system, this system might be a temporal variable with uncertainties and nonlinearities caused by the turbine. Hence, control algorithms appropriate for these structures are required when designing a supplementary controller for the load frequency problem. MATLAB snippets are listed below.

Initialization

```
function [X,V,X_Pbest,Pbest,Gbest]=Initialize
(Population_Max,XMax)

for C1=1:Population_Max
    X(C1,:)=rand(1,3)*XMax;
    V(C1,:)=rand(1,3)*XMax;
    Pbest(C1)=0;
end

X_Pbest=X;
Gbest=0;
```

Main MATLAB snippet

```
%%%%%%%%%%%%%%%%%%%%%%%%%%%%%%%%%%%%%%%%%%%%%%%%%%%%%%%%%%%%%%%%%
%   Multiobjective With PSO : PID
%%%%%%%%%%%%%%%%%%%%%%%%%%%%%%%%%%%%%%%%%%%%%%%%%%%%%%%%%%%%%%%%%

clc;
clear;
close all;
load ('WPoint.txt');
WPoint=WPoint(15,:);
wew=25;
Population_Max=30;
Max_Generation = wew;
Extract=20;
XMax=3;
c1=1;
c2=1;
OTime=1000;

rand ('state',2);

[X,V,X_Pbest,Pbest,Gbest]=Initialize (Population_Max,XMax);

Archive=[];
Temp=hardlims (rand(1,Population_Max)*2-1)';
for i=1:3
    Direction(1:Population_Max,i)=Temp;
end
Total=zeros (Population_Max,1);

%%%%%%%%%%%%%%%%%%%%%%%%%%%%%%%%%%%%%%%%%%%%%%%%%%%%%%%%%%%%%%%%%
% Start
%%%%%%%%%%%%%%%%%%%%%%%%%%%%%%%%%%%%%%%%%%%%%%%%%%%%%%%%%%%%%%%%%

for Generation=1:Max_Generation

    WT=randint;
```

```
    if WT==0
        l1=rand;
        l2=rand;
        W1=l1/(l1+l2);
        W2=l2/(l1+l2);
    else
        W1=1;
        W2=1;
    end

    for Chromosome=1:Population_Max

        for C1=1:3
            if X(Chromosome,C1)>XMax
                X(Chromosome,C1)=rand*XMax;
            end
            if X(Chromosome,C1)<0
                X(Chromosome,C1)=rand*XMax;
            end
        end

        [O1,O2,T1,T2,G1,G2,E1,E2]=Sim_PID2
 (WPoint,X(Chromosome,:),OTime);

        if T1<200
            T1=OTime*5;
        end
        if T2<200
            T2=OTime*5;
        end

        F2(Chromosome,1)=1/(1+(O1+O2)/2);
        F1(Chromosome,1)=OTime/(OTime+(T1+T2)/2);

        Eval(Chromosome,1)=W1*F1(Chromosome)+W2*F2(Chromosome);

        if Eval(Chromosome)>=Gbest
            Gbest=Eval(Chromosome);
            X_Gbest=X(Chromosome,:);
        end

        if Eval(Chromosome)>=Pbest(Chromosome)
            Pbest(Chromosome)=Eval(Chromosome);
            X_Pbest(Chromosome,:)=X(Chromosome,:);
        end

    end

    Alpha=0.5+Generation/Max_Generation;
    Inertia=0.25+rand*0.75;

    for Chromosome=1:Population_Max
        r1=rand;
        r2=rand;
        R1=r1*c1;
        R2=r2*c2;


V_New(Chromosome,:)=Inertia*V(Chromosome,:)+Alpha*(R1*(X_Pbest(Chro
mosome,:)-X(Chromosome,:))+R2*(X_Gbest-X(Chromosome,:)));
```

```
        for C1=1:length(V_New(Chromosome,:))
            if V_New(Chromosome,C1)<-XMax
                V_New(Chromosome,C1)=-XMax;
            end
            if V_New(Chromosome,C1)>XMax
                V_New(Chromosome,C1)=XMax;
            end
        end

        X_New(Chromosome,:)=X(Chromosome,:)+V_New(Chromosome,:);

        for C1=1:3
            if X_New(Chromosome,C1)>XMax
                X_New(Chromosome,C1)=rand*XMax;
            end
            if X_New(Chromosome,C1)<0
                X_New(Chromosome,C1)=rand*XMax;
            end
        end

        [O1,O2,T1,T2,G1,G2,E1,E2]=Sim_PID2
(WPoint,X_New(Chromosome,:),OTime);

        if T1<200
            T1=OTime*5;
        end
        if T2<200
            T2=OTime*5;
        end

        F2_New(Chromosome,1)=1/(1+(O1+O2)/2);
        F1_New(Chromosome,1)=OTime/(OTime+(T1+T2)/2);

    end

    for m=1:1
        p1=randint (1,1,[1,Population_Max]);
        p2=randint (1,1,[1,Population_Max]);

        X1T=X(p1,:);
        X2T=X(p2,:);
        V1T=V(p1,:);
        V2T=V(p2,:);

        r1=rand;
%         r2= 8-r2;
        r2=rand;

        X(p1,:)=r1*X1T+r2*X2T;
        X(p2,:)=r1*X2T+r2*X1T;
        V(p1,:)=r1*V1T+r2*V2T;
        V(p2,:)=r1*V2T+r2*V1T;

        X_Pbest (p1,:)=X(p1,:);
        X_Pbest (p2,:)=X(p2,:);

        [O1,O2,T1,T2,G1,G2,E1,E2]=Sim_PID2 (WPoint,X(p1,:),OTime);
        F1(p2,1)=OTime/(OTime+(T1+T2)/2);
        F2(p1,1)=1/(1+(O1+O2)/2);

        [O1,O2,T1,T2,G1,G2,E1,E2]=Sim_PID2 (WPoint,X(p2,:),OTime);
```

```
        F1(p2,1)=OTime/(OTime+(T1+T2)/2);
        F2(p2,1)=1/(1+(O1+O2)/2);

    end

    Temp_List=[F1,F2,X,V;F1_New,F2_New,X_New,V_New];
    [X,V,F1,F2]=Sort_List (Temp_List,Population_Max);

    figure (1);
    clf;
    Temp1=(OTime-OTime*F1(:,1))./F1(:,1);
    Temp2=( 8-1*F2(:,1))./F2(:,1);
    plot (Temp1,Temp2,'ro');
    hold on;
    title (int2str(Generation));
    xlabel ('St Time');
    ylabel ('Overshoot');
    pause (0.01);

    if Generation>wew-1 && mod (Generation,1)==0
        Temp_List=[F1,F2,X];
        Archive=[Archive;Temp_List];
    end

    if Generation>1
        Old_SD=New_SD;
        New_SD=Pbest'+0.5*((F1+F2));
        Temp=hardlims(Old_SD-New_SD);
        Total=Total+Temp;
        I=3;
        for i=1:Population_Max
            if Total(i)>0
                Total(i)=0;
            end
            if Total(i)<-I
                Direction(i,:)=-Direction(i,:);
                Total(i)=0;
            end
        end
    else
        New_SD=Pbest'+0.5*((F1+F2));
    end

    if sum(abs(V(:)))<25
%           X=X+Direction*rand*0.01;
        X=X+Direction*rand*0.05/Generation;
    end

%%%%%%%%%%%%%%%%%%%%%%%%%%%%%%%%%%%%%%%%%%%%%%%%%%%%%%%%%%%%%%%%%%%%%%%%
end  % End of generation loop
%%%%%%%%%%%%%%%%%%%%%%%%%%%%%%%%%%%%%%%%%%%%%%%%%%%%%%%%%%%%%%%%%%%%%%%%

% Archive=Correct_Archive (Archive,0.85,WPoint);

% List=Extract_Archive (Archive,Extract,0.5,0.5);
List=Make_Archive (Archive);
List=sortrows (List,1);
List=Prun_Archive (List);
Show_Result (List(:,1),List(:,2),OTime*.01);
X_Best=List(:,3:5);
```

```
for i=1:length (X_Best(:,1))
    [O1,O2,T1,T2,G1,G2,E1,E2]=Sim_PID2 (WPoint,X_Best(i,:),1000);

    figure (2);
%     clf;
if (T1+T2)/2 <850
    plot (G2);
end
    hold on;
%     plot (G2,'r');
    pause (1);
end

%%%%%%%%%%%%%%%%%%%%%%%%%%%%%%%%%%%%%%%%%%%%%%%%%%%%%%%%%%%%%%%%%%
% End
%%%%%%%%%%%%%%%%%%%%%%%%%%%%%%%%%%%%%%%%%%%%%%%%%%%%%%%%%%%%%%%%%
```

Data archiving snippet

```
function [List]=Make_Archive (Temp_List)

F=[Temp_List(:,1),Temp_List(:,2)];
DC=1;                                   % Dominate Counter
NDC=1;

for C1=1:length(Temp_List(:,1))

    C2=1;                               % Terminator Tester
    D_Test=0;                           % Dominate Tester
    while C2<=length(Temp_List(:,1)) && D_Test==0
        if (F(C1,1)>F(C2,1) && F(C1,2)>F(C2,2)) ||
(F(C1,1)>F(C2,1) && F(C1,2)==F(C2,2)) || (F(C1,1)==F(C2,1) &&
F(C1,2)>F(C2,2))
%           if F(C1,1)>F(C2,1) && F(C1,2)>F(C2,2)
              D_Test=1;
        end
        C2=C2+1;
    end

    if D_Test==0
        Non_Dominate (NDC,:)=Temp_List(C1,:);
        NDC=NDC+1;
    else
        Dominate (DC,:)=Temp_List(C1,:);
        DC=DC+1;
    end
end

List=Non_Dominate;
```

calling

```
function [Archive]=Prun_Archive (Archive)

if length (Archive (:,1))>1
    j=2;
    Temp(1,:)=Archive (1,:);
    for i=2:length (Archive (:,1))
        if Archive (i,1)==Temp (j-1,1) && Archive (i,2)==Temp (j-
1,2)

        else
            Temp (j,:)=Archive(i,:);
            j=j+1;
        end
    end
    Archive=Temp;
end
```

Showing results

```
function Show_Result(F1,F2,OTime)

Figure=figure(1);
set (Figure,'NumberTitle','off');
set (Figure,'Name','Final Pareto Front');
clf;
Temp1=(OTime-OTime*F1(:,1))./F1(:,1);
Temp2=( 8-1*F2(:,1))./F2(:,1);
plot (Temp1(:,1),Temp2(:,1),'bo');
hold on;
legend ('Pareto Front');
xlabel('settling time');
ylabel('overshoot/undershoot');
```

PID snippet

```
function [O1,O2,T1,T2,G1,G2,E1,E2]=Sim_PID2 (WPoint,PID,Time)
Tg1=0.08;
kp1=120;
Tg2=.08;
kp2=120;
R1=2.4;
R2=2.4;
Tt1=0.3;
Tt2=0.3;
%=================\
cp11=.25 ;cp12=.0  ;cp13=.25    ;cp14=.25;
cp21=.25 ;cp22=.25  ;cp23=.25    ;cp24=.0;
cp31=0  ;cp32=.5 ;cp33=.5    ;cp34=.5;
cp41=.5 ;cp42=.25  ;cp43=.0   ;cp44=.25;

GPM=[cp11 cp12 cp13 cp14
    cp21 cp22 cp23 cp24
    cp31 cp32 cp33 cp34
    cp41 cp42 cp43 cp44];
% All load Demanded by each Discos& AREAs
num_disco=4;
num_genco=4;
num_area=2;
DP1=[.1 .1 .1 .1];%the rows of this vector are demanded of discos
from area.

for i=1:num_disco
    DPlg1(i)=GPM(1,i)*DP1(1,i);%load of genco first.
    DPlg2(i)=GPM(2,i)*DP1(1,i);%load of genco first.
    DPlg3(i)=GPM(3,i)*DP1(1,i);%load of genco first.
    DPlg4(i)=GPM(4,i)*DP1(1,i);%load of genco first.
```

```
end
DPref1=sum(DPlg1);
DPref2=sum(DPlg2);
DPref3=sum(DPlg3);
DPref4=sum(DPlg4);

DPd1=.1;%the uncontracted load in 1st area
DPd2=.200;%the uncontracted load in 2end area

%compute the all load in each area
DPL1=DPl(1,1)+DPl(1,2)+DPd1;
DPL2=DPl(1,3)+DPl(1,4)+DPd2;
%8888888888888888888888888888888888888888888
Tp1=WPoint(1);
Tp2=Tp1;
Tp3=Tp2;
Tp4=Tp3;

T12=WPoint(2);

B1=WPoint(3);
B2=B1;
B3=B2;
B4=B2;
Tg1=0.08;
Tg2=.08;
Tg3=Tg2;
Tg4=Tg2;

kp1=120;
kp2=120;
kp3=kp2;
kp4=kp2;

R1=2.4;
R2=2.4;
R3=R2;
R4=R2;

Tt1=.3;
Tt2=.3;
Tt3=Tt2;
Tt4=Tt2;

Df1=[0 0];
Df2=Df1;


Dpti=[.01];
Dxg1=Df1;
Dxg2=[0 0];
Dxg3=[0 0];
Dxg4=[0 0];

Dpg1=Df1;
Dpg2=Df1;
Dpg3=Df1;
Dpg4=Df1;

upi_1=[0 0];
upi_2=[0 0];
```

```
u_1=[0 0];
u_2=[0 0];

Ace1=[0 0];
Ace2=[0 0];


Aa=[.5 .5 .5 .5];
Tsam=0.01;


Df1=[0 0];
Df2=[0 0];
Dpti=[0 0];
Dxg1=[0 0];
Dxg2=[0 0];
Dpg1=[0 0];
Dpg2=[0 0];
upi_1=[0 0];
upi_2=[0 0];
u_1=[0 0];
u_2=[0 0];
Ace1=[0 0];
Ace2=[0 0];

KP1=PID(1);
KP2=KP1;

KI1=PID(2);
KI2=KI1;

KD1=PID(3);
KD2=KD1;
u(2)=0;
O2=0;

Df(1)=0;
u_2(2)=0;
u_1(2)=0;
u_3(2)=0;
u_4(2)=0;

O2=0;
Df(1)=0;
Ace2(2)=0;
Ace1(2)=0;

Dpti_sched_eport1=sum((cp13+cp23)*DPl(1,3)+(cp14+cp24)*DPl(1,4));
Dpti_sched_import1=sum((cp31+cp41)*DPl(1,1)+(cp32+cp42)*DPl(1,2));
Dpti_sched1=Dpti_sched_eport 8-Dpti_sched_import1;
Dpti_sched2=-Dpti_sched1;

for k1=2:Time-1

    Df1(k1+1)=Df1(k1)   +Tsam*(1/Tp1)*(-
Df1(k1)+kp1*(Dpg1(k1)+Dpg2(k1))-kp1*Dpti(k1)-kp1*DPL1);
    Df2(k1+1)=Df2(k1)   +Tsam*(1/Tp2)*(-
Df2(k1)+kp2*(Dpg3(k1)+Dpg4(k1))+kp2*Dpti(k1)-kp2*DPL2);

    Dpg1(k1+1)=Dpg1(k1)+Tsam*(1/Tt1)*(-Dpg1(k1)+Dxg1(k1));
```

```
    Dpg2(k1+1)=Dpg2(k1)+Tsam*(1/Tt2)*[-Dpg2(k1)+Dxg2(k1)];
    Dpg3(k1+1)=Dpg3(k1)+Tsam*(1/Tt3)*(-Dpg3(k1)+Dxg3(k1));
    Dpg4(k1+1)=Dpg4(k1)+Tsam*(1/Tt4)*[-Dpg4(k1)+Dxg4(k1)];

    Dxg1(k1+1)=Dxg1(k1)+Tsam*(1/Tg1)*((-1/R1)*Df1(k1)-
Dxg1(k1)+DPref1+u_1(k1));
    Dxg2(k1+1)=Dxg2(k1)+Tsam*(1/Tg2)*((-1/R2)*Df1(k1)-
Dxg2(k1)+DPref2+u_2(k1));
    Dxg3(k1+1)=Dxg3(k1)+Tsam*(1/Tg3)*((-1/R3)*Df2(k1)-
Dxg3(k1)+DPref3+u_3(k1));
    Dxg4(k1+1)=Dxg4(k1)+Tsam*(1/Tg4)*((-1/R4)*Df2(k1)-
Dxg4(k1)+DPref4+u_4(k1));

    Dpti(k1+1)=Dpti(k1)+Tsam*T12*(Df1(k1)-Df2(k1));
    Dpti_erro1(k1+1)=Dpti(k1+1)-Dpti_sched1;
    Dpti_erro2(k1+1)=-Dpti_erro1(k1+1);

    Ace1(k1+1)=B1*Df1(k1+1)+Dpti_erro1(k1+1);
    Ace2(k1+1)=B2*Df2(k1+1)-Dpti_erro1(k1+1);

    y_1(k1)=Ace1(k1+1);
    y_2(k1)=Ace2(k1+1);

    upi_1(k1+1)=upi_1(k1)-((KD1/Tsam)*Ace1(k 8-1)-
(KP1+2*(KD1/Tsam))*Ace1(k1)+((KD1/Tsam)+KI1+KP1)*Ace1(k1+1));

    upi_2(k1+1)=upi_2(k1)-((KD2/Tsam)*Ace2(k 8-1)-
(KP2+2*(KD2/Tsam))*Ace2(k1)+((KD2/Tsam)+KI2+KP2)*Ace2(k1+1));

    U_1(k1+1)=upi_1(k1+1);

    U_2(k1+1)=upi_2(k1+1);

    u_1(k1+1)=Aa(1,1)*(U_1(k1+1));
    u_2(k1+1)=Aa(1,2)*(U_1(k1+1));

    u_3(k1+1)=Aa(1,3)*(U_2(k1+1));
    u_4(k1+1)=Aa(1,4)*(U_2(k1+1));

end

E1=Dpti;
E2=[Ace1;Ace2;Dpg3;Dpg4];

G1=Df1;
G2=Df2;

Ace1=abs(Ace1);
Ace1N=(Ace 8-min(Ace1))/(max(Ace1)-min(Ace1));
for i=3:length(Ace1N)-50
    if sum (abs(Ace1N(i:i+50)))<0.01
        break
    end
end
T1=i;
O1=max(Ace1);

Ace2=abs(Ace2);
Ace2N=(Ace2-min(Ace2))/(max(Ace2)-min(Ace2));
for i=3:length(Ace2N)-50
    if sum (abs(Ace2N(i:i+50)))<0.01
```

```
        break
    end
end
T2=i;
O2=max(Ace2);
```

Data short list function

```
function [X,V,F1,F2]=Sort_List (Temp_List,Population_Max)

NDC=1;                                          % Non-Dominate Counter
while NDC<=Population_Max

    F=[Temp_List(:,1),Temp_List(:,2)];
    DC=1;                                       % Dominate Counter
    clear Dominate;

    for C1=1:length(Temp_List(:,1))

        C2=1;                                   % Terminator Tester
        D_Test=0;                               % Dominate Tester
        while C2<=length(Temp_List(:,1)) && D_Test==0
            if (F(C1,1)<F(C2,1) && F(C1,2)<F(C2,2)) ||
(F(C1,1)<F(C2,1) && F(C1,2)==F(C2,2)) || (F(C1,1)==F(C2,1) &&
F(C1,2)<F(C2,2))
%               if F(C1,1)>F(C2,1) && F(C1,2)>F(C2,2)
                D_Test=1;
            end
            C2=C2+1;
        end

        if D_Test==0 && NDC<=Population_Max
            Non_Dominate (NDC,:)=Temp_List(C1,:);
            NDC=NDC+1;
        else
            Dominate (DC,:)=Temp_List(C1,:);
            DC=DC+1;
        end
    end
    Temp_List=Dominate;
end

X=Non_Dominate (:,3:5);
V=Non_Dominate (:,6:8);
F1=Non_Dominate (:,1);
F2=Non_Dominate (:,2);
```

Uncited Reference

1. H.B. Demuth, M.H. Beale, O. De Jess, M. T. Hagan, *Neural network design* (Martin Hagan, 2014)

Correction to: Neural Networks and Learning Algorithms in MATLAB

Correction to:
A. Mohammadzadeh et al., *Neural Networks and Learning Algorithms in MATLAB*, Synthesis Lectures on Intelligent Technologies,
https://doi.org/10.1007/978-3-031-14571-1

In the original version of the chapter, the following belated corrections have been incorporated: The editor name "Ardahir Mohammadazadeh" has been changed to "Ardashir Mohammadzadeh" in the Frontmatter and in Chapters 1–8. The correction/erratum chapter and the book have been updated with the change.

The updated version of the book can be found at
https://doi.org/10.1007/978-3-031-14571-1

A. Mohammadzadeh et al., *Neural Networks and Learning Algorithms in MATLAB*,
Synthesis Lectures on Intelligent Technologies,
https://doi.org/10.1007/978-3-031-14571-1_9

MIX
Papier aus verantwortungsvollen Quellen
Paper from responsible sources
FSC® C105338

If you have any concerns about our products,
you can contact us on
ProductSafety@springernature.com

In case Publisher is established outside the EU,
the EU authorized representative is:
Springer Nature Customer Service Center GmbH
Europaplatz 3, 69115 Heidelberg, Germany

Printed by Libri Plureos GmbH
in Hamburg, Germany